D0586185

UNIVERSITY
NEWCASTLE
UPON TYNE
LIBRARY

Effective Management of Benchmarking Projects

Practical Guidelines and Examples of Best Practice

Mohamed Zairi BSc., MSc., Postgrad.Dipl., PhD, FRSA

Sabic Professor of Best Practice Management, University of Bradford Management Centre

OXFORD BOSTON JOHANNESBURG MELBOURNE NEW DELHI SINGAPORE

Butterworth-Heinemann
Linacre House, Jordan Hill, Oxford OX2 8DP
225 Wildwood Avenue, Woburn, MA 01801-2041
A division of Reed Educational and Professional Publishing Ltd

 A member of the Reed Elsevier plc group

First published 1998

British Library Cataloguing in Publication Data
A catalogue record for this book is available from the British Library

Library of Congress Cataloguing in Publication Data
A catalogue record for this book is available from the Library of Congress

ISBN 0 7506 3987 3

Typeset by Wyvern 21 Ltd, Bristol
Printed and bound in Great Britain by
Biddles Ltd, Guildford and King's Lynn

Contents

Acknowledgements

I am grateful to so many individuals and organizations with whom I have had the privilege to work, to share ideas with, and who have helped shape my preliminary ideas, clarify my confusion over certain things and enrich my academic knowledge with application-based learning. The mission of a person like myself is to help disseminate learning and to ensure that society at large benefits from progress and advancement in all corners of the globe. This effort therefore sounds very daunting and very challenging indeed. It is, however, often made easy because there is a **benchmarking community** out there! There are many advocates and 'gurus' who work tirelessly to share ideas, experiment with new learning and facilitate the introduction of change in a variety of organizational contexts.

To all of you men and women out there, I would like to dedicate this book to you and I hope that the collection of ideas put together can help enhance your own knowledge base and can provide you with further impetus to carry on benchmarking.

As usual, my sincere gratitude must also go to organizations that I am so proud of, that are truly world-class in what they do and who have always been so keen to share their learning.

Dedication

To my wife Alweena

Thank you for your unwavering support, for accepting to share my vision and destiny and for enduring all the pain, pressure and sacrifices

and

To my children Adel, Bilal and Nadir

Thank you for showing so much maturity and understanding to a very often 'distant father'

Introduction

> Consumers, by seeking quality and value, set the standards of acceptability for products and services by voting with their market-place dollars
>
> *Ronald Reagan*

> Our social mission as a manufacturer is only realized when products reach, are used by, and satisfy the customer . . . We need to take the customer's skin temperature daily
>
> *Konosuke Matsushita*

> High quality means pleasing customers, not just protecting them from annoyances
>
> *David Garvin*

This book needed to be written since there are still many burning issues associated with the application of benchmarking which remain to be addressed. The market is flooded with textbooks on the *theory* of benchmarking. Nearly all the texts written so far are similar in their approach. They tend to describe:

- What benchmarking is
- Its origins
- The benchmarking process covered step by step.

The two major issues that have not been addressed effectively relate to:

- How can benchmarking be applied in a variety of scenarios?
- Where are *real* examples of effective applications?

Effective project management seeks to address both areas and therefore does represent a *first* in this field and will, without doubt, fulfil a role and satisfy a current need in the marketplace. It will also represent the *ideal companion* to best practice benchmarking. The former is a complete and comprehensive guide for the concept of benchmarking, applications and examples of best practice. The latter seeks to focus on all aspects of project management by explicitly covering methodological issues (associated with partner selection, for instance), prioritization, managing industrial visits, among others.

It is gratifying and reassuring to see that the art of benchmarking has significantly evolved and is no longer referred to as a mechanistic tool based on *Phases* and a series of *Steps*. Karlof and Ostblom, for instance, have coined the term *benchlearning* [1]. They define the latter as '. . . creating an environment which rewards constant learning with better performance and, in consequence, with greater success'.

Indeed, the essence of benchmarking is to encourage continuous learning and to lift organizations to higher competitive levels. Through problem-solving, the acquisition of internal and external knowledge and its effective implementation, standards of practice can be enhanced with the direct effect of achieving higher levels of customer satisfaction and, as a consequence, business performance can also be greatly improved. As the saying goes:

> As the soil cannot be productive without culture, so the mind without cultivation can never produce good fruit. (Seneca)

Organizational excellence can only come through a continuous process of replenishing the mind with new ideas, good practices and a progressive way to tackle problems and work innovatively and creatively for the end benefit of the customer. As the other saying also goes: 'antiquity is not a voucher for efficiency'.

Doing more of the same can only give more of the same. It is true that the battle of the 1990s and beyond is the battle for branding, differentiation, customer loyalty and retention. The true differentials of effective performance are those which are going to be defined by constant innovation and creativity, a knowledge-based approach to competitiveness and a desire to win the hearts and the minds of customers through uniqueness, quality, and valued service and customer focus.

The literature is rife with scepticism on the workings of benchmarking and reports on failed projects. Very often it is found that organizations embark on benchmarking expeditions without:

- Necessary preparation
- Poor training and understanding of the process of benchmarking
- Unmanaged and uncontrolled processes
- Lack of clarity on which are the specific areas to be benchmarked
- Not linking the areas selected to strategic imperatives
- Poor partner selection
- No follow-up from company visits
- Not linking the benchmarking outcomes to areas critically concerned
- Limiting the exercise of benchmarking to number gathering and the building of comparative data

- Not appreciating that fundamentally benchmarking has to be linked to Total Quality Management, processes and customers.

It is obvious therefore that all the reported instances of malpractice and misuse and abuse of benchmarking are due to a wide variety of reasons, none of them, however, indicating that the 'tool' itself does not work, is a 'fad' or is of great limitation. The inability of organizations to quantify the benefits from benchmarking expeditions does reside in the hands of senior managers who do not really appreciate the aspects of benchmarking, for:

> Learning is a dangerous weapon and apt to wound its master if it be wielded by a feeble hand. (Montaigne)

Effective benchmarking project management intends to equip project teams and senior managers with all the necessary competence for managing projects effectively. We begin the book with early definitions of 'what to benchmark' and end with simulating a real case study where a benchmarking project was conducted by observing all the necessary rules and with total adherence to the various protocols.

The structure of the book was developed to cover nine distinctive areas:

1 **Rank Xerox: where the benchmarking story started**. This chapter addresses the story of benchmarking within Rank Xerox and its link with total quality management. The chapter also contains examples of benchmarking applications.
2 **The strategic application of benchmarking for best practice: a prioritization tool**. Very often organizations don't know where to start with benchmarking, which areas to choose and how they can be sure that the process selected affects business performance significantly. This chapter deals with this issue and presents a methodology together with an example of a real application.
3 **Partner selection**. A critical area of benchmarking is whom to learn from and what is best practice. Can other companies help us? This chapter addresses the issue of partnerships in benchmarking as well as others.
4 **The ethics of benchmarking: how to handle secrecy and other sensitive issues**. Benchmarking is sometimes mistaken for industrial espionage and it is often the case that many organizations do not indulge in benchmarking because of fear of harm to their businesses. This chapter is therefore very critical in dispelling the myth about sensitivities and industrial espionage and will also cover the 'rules of the game' and how to engage in benchmarking activity.
5 **The value of industrial visits**. Industrial visits have in some cynical quarters been described as an opportunity for 'industrial tourism'. This chapter

puts industrial visits in the right context and describes the whole process of managing them with illustrations and examples of best practice.

6 **The value of benchmarking awards**. This chapter describes the value of benchmarking awards and one well-established process and comprehensively covers two case studies of winners of the European Best Practice Benchmarking Award.
7 **The process of benchmarking in practice**. The generic methodology of benchmarking is well documented in all the textbooks. This chapter specifically looks at the approach of benchmarking adopted by several organizations and how they attempt to develop a culture of continuous learning.
8 **The diversity of benchmarking: examples of best practice**. This chapter discusses the application of benchmarking in a wide variety of contexts, covering public and private sector industries. It represents an opportunity to highlight that benchmarking is a multi-faceted process and can be applied in a multitude of situations.
9 **Bringing it all together; effective project management**. Using one particular application of benchmarking, this chapter discusses the ideal project management aspects for a successful application.

Reference

1 Karlof, B. and Ostblom, S. (1993) *Benchmarking: A Signpost to Excellence in Quality and Productivity*, John Wiley, New York.

1 Rank Xerox: where the benchmarking story started

Between knowledge of what really exists and ignorance of what does not exist lies the domain of opinion. It is more obscure than knowledge, but clearer than ignorance

Plato

As a general rule in the most successful man in life is the man who has the best information

Benjamin Disraeli

To most men, experience is like the stern lights of a ship, which illuminate only the track it has passed

S. T. Coleridge

1.1 The quality journey within Rank Xerox

Rank Xerox was a superior competitor since its foundation in 1956. However, there was a sharp erosion in its level of competitiveness to the benefit of the Japanese who managed to capture the market with high-quality, small copiers at a low cost.

The drive for quality became necessary in the early 1980s as a result of a sharp decrease in growth in revenue and profit. In 1980, for instance, ROA was 19 per cent but this had declined to 8.4 per cent by 1983. In Europe 80 per cent of the copiers in use were Japanese by the mid-1980s. This, of course, was a major drop from the days when Xerox was in a commanding position, when it grew from a small company with revenues of $33 million in 1959 to a major corporation with revenues of $176 million by 1963 and $4 billion by 1975. Xerox, of course, took advantage of the favourable situation in the market-place. It was protected by its patented reprographic processes then, with very little competition. As Barry Rand (President, United States Marketing Group) puts it: '. . . It's not hard to be the best player when you're the only player'. Xerox's success was also spurred by the fact that there was a lot of demand for plain-paper copying. In the process of trying to capitalize on the available opportunities and satisfy existing levels of demand, Xerox lost sight of the customer, thus giving the competition the chance to innovate quickly and effectively and to start to challenge Xerox's market share. Realizing the gravity of the situation, Xerox decided to take action. Nobody under-estimated the magnitude of the task for a corporation the size of Xerox ($9 billion) and with more than 100 000 staff:

1 Xerox therefore launched a company-wide process called 'Business Effectiveness' based on a combination of employee involvement management style and a competitive process.
2 More attention was also given to obstacles to quality and areas leading to high cost by focusing on all aspects of product design, planning, engineering and production.
3 The customer was refocused through the development of a customer-satisfaction measurement system.
4 Much effort was devoted to ensuring that Xerox's new products were of the highest quality and that they would be able to assist Xerox to reclaim its market share.

These changes were very effective and led to significant progress.

1.2 Leadership through quality: a total quality process for Xerox

The changes were put in place with a high degree of effectiveness, and included, among others:

- A change in the approach to product development and delivery
- A change in the way Xerox approached its customers
- A change in Xerox's cost base
- Employee involvement programmes
- Introduction of competitive benchmarking.

The senior management team realized that all the above processes were working well only on some levels and tended to concentrate on uncoordinated and isolated areas. They decided that there was a need for a unifying process which would ensure employee involvement at all levels. These reinforcements came from the following four conclusions:

1 The Japanese success was due to the fact that their management process was driven by their commitment to total quality management.
2 Fuji Xerox in Japan had demonstrated the power of quality through achieving excellent business results and winning the ultimate quality prize (Deming Prize).

3 Quality started to spread in America and Europe, and many CEOs started to build it with their competitive objectives.
4 Through a strategic business analysis Xerox realized it could grow to a $25–30 billion corporation by 1992 through the use of quality management principles.

This led to the birth of 'leadership through quality' (LTQ).

1.2.1 What is leadership through quality?

LTQ is presented as an integrated philosophy with the following key areas of focus:

- A *goal* for Xerox to attain and maintain
- A *strategy* to enable Xerox to achieve its competitive advantage
- A *way of working* or *process* to use for managing operation of the business, and at all levels.

Leadership through quality works through five major mechanisms:

1 Standards and measurement:

   - A six-step problem-solving process
   - A nine-step quality improvement process
   - Competitive benchmarking.

   All tools must place emphasis on doing the right things right first time and determining the cost of quality.
2 Recognition and reward: recognizing quality efforts and contributions to continuous improvement in various forms.
3 Communications: formally and informally people are told about targets and objectives of the corporate organization, and are given feedback on a regular basis.
4 Training: training *all* employees in the principles of leadership through quality and giving them a good knowledge of the workings of the various problem-solving and continuous-improvement tools to use in their immediate work areas.
5 Management behaviours and actions: managers must adhere to the principles of leadership through quality and practise them in their everyday job. In other words, managers must walk like they talk. As expressed in the 1983 document which launched LTQ: 'The introduction and adoption of leadership through quality will be fostered, encouraged and led by management at all levels.'

1.2.2 How does leadership through quality work?

The tools which are used to ensure that Rank Xerox does the right things right first time and all the time include:

- The problem-solving process
- The quality-improvement process
- The benchmarking process
- The self-assessment process (Business Excellence Certification Model.)

Here we will examine the first two tools and the last two will be discussed later.

1. **The problem-solving process**: This problem has six steps and is illustrated in Figure 1.1. It is to enable people to close gaps in performance and to analyse problems, develop solutions and put action plans together.

    - Identifying and selecting a problem: using brainstorming techniques to define opportunities and priorities for improvement.
    - Analysing the problem: this stage requires the collection of data and its analysis, using simple tools of quality improvement and other statistical techniques.
    - Figure 1.2 illustrates an example of one of the powerful tools used, a cause-and-effect diagram. Other tools which can be used in the data collection and analysis stage may include:

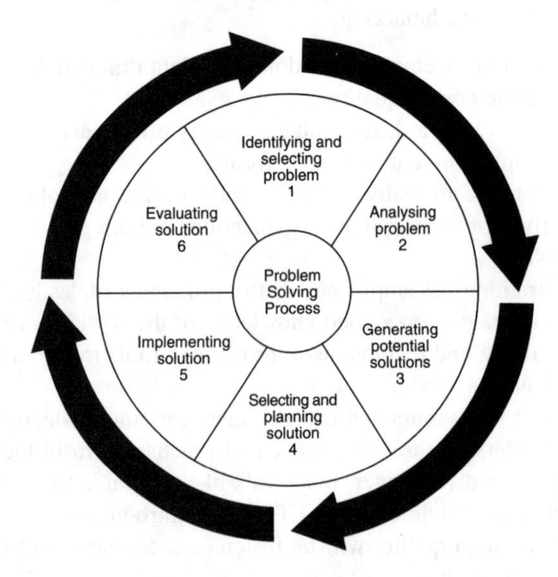

Figure 1.1 *The problem-solving process used at Rank Xerox*

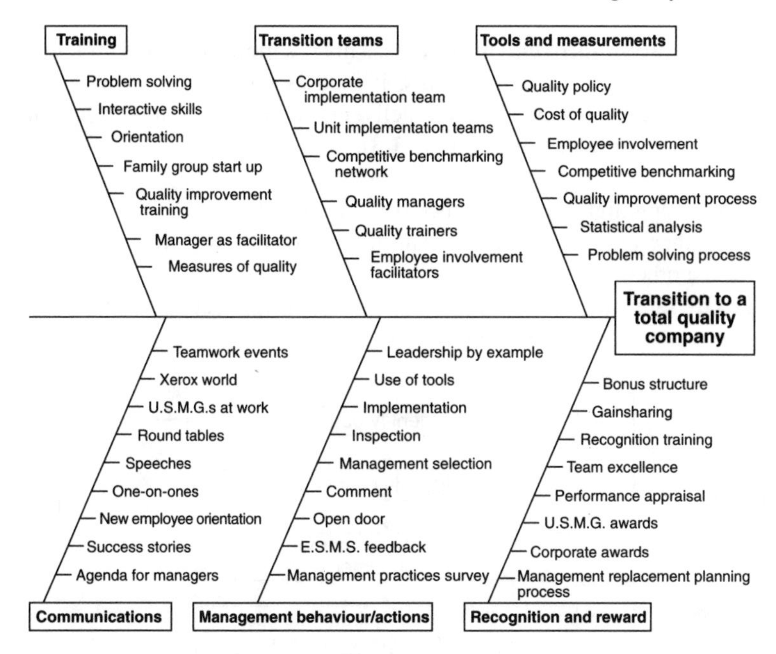

Figure 1.2 *Cause-and-effect diagram application of Rank Xerox*

- Force-field analysis
- Histograms
- Pareto charts
- Pie charts
- Time charts
- Flow charts
- Control charts
- Cost–benefit analysis

- Generating potential solutions: potential solutions and various alternatives may be recommended at this stage, using brainstorming techniques.
- Selecting and planning a solution: the teams concerned are expected to choose the most feasible and practical solution and to put together an implementation plan for it with the full support of those most involved.
- Implementing the solution: to go ahead with implementation after management approval and fine-tune and steer the change positively forward to achieve the desired effect.
- Evaluating the solution: through measurement and feedback the teams concerned can determine whether the implemented solution has had the desired effect and if not they can repeat the cycle and continue monitoring.

2 **The quality-improvement process**. This is a more pervasive tool, it is not related only to internal *problems*, it focuses more on *process* routes for products and services which are delivered to the end customer. It is therefore customer related (see Figure 1.3).

Table 1.1 highlights the difference between the PSP and the QIP. Very often, however, they are complementary and the use of one will trigger the utilization of the other.

The QIP illustrated in Figure 1.3 has nine steps which are grouped into planning, organizing and monitoring stages:

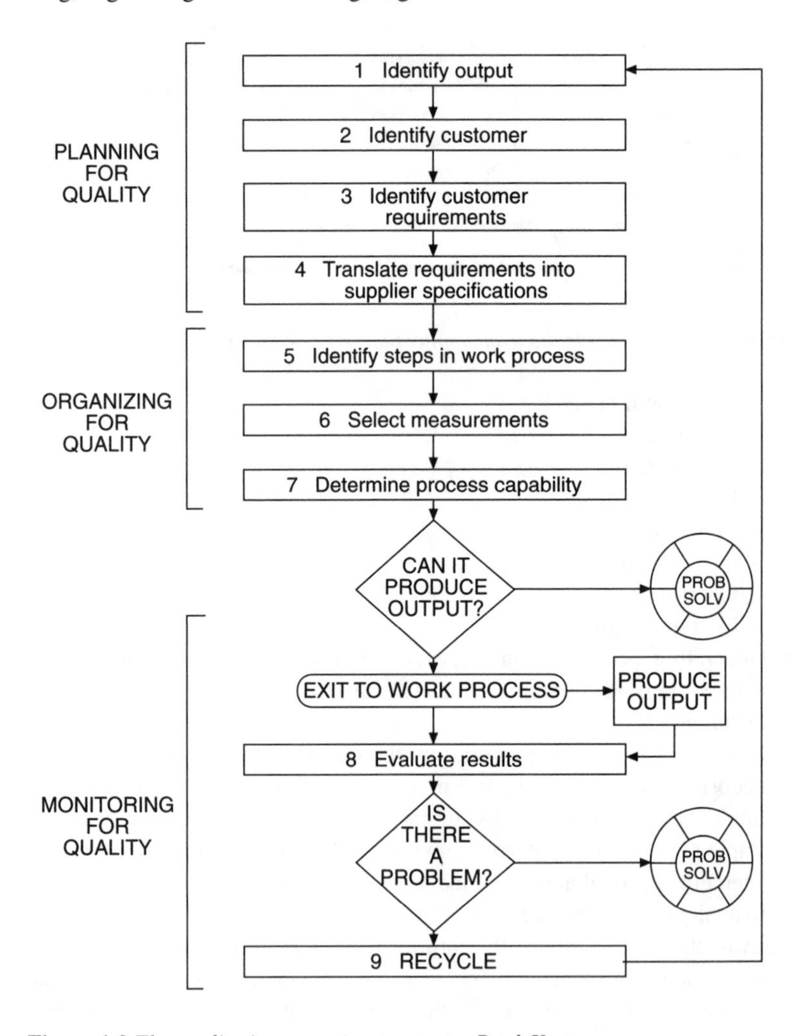

Figure 1.3 *The quality-improvement process at Rank Xerox*

Table 1.1 *The differences between the problem-solving and the quality-improvement processes at Rank Xerox*

Use problem solving when	*Use quality improvement when*
There is a gap between what is happening and what you want	You need to improve the quality of a particular, currently existing output
You want to move from a vague dissatisfaction to a solvable, clearly defined problem	You don't have agreed-upon customer requirements for an output
You're not sure how to approach an issue	You are about to produce a new output

1. Identify output: the team to brainstorm and define the desired output.
2. Identify customer: this refers to customers of the desired output, the outcome of using the QIP (often internal customers).
3. Identify customer requirements: this stems from stage 2 and will prompt the project team (suppliers) to work closely with customers (beneficiaries of the output) to define what is required and therefore how it is going to be delivered.
4. Translate requirements into supplier specification: all requirements are put into measurable and achievable deliverables.
5. Identify steps in the work process: a step-by-step approach to how the output is going to be produced needs to be developed, using perhaps existing work procedures and guidelines and producing a flow chart.
6. Select measurement: measures need to be selected to assess before, during and after scenarios and also to design measures for continuous monitoring and prevention purposes.
7. Determine process capability: this to test the recommended process and to ensure that it can do the right things right first time. Otherwise the team can use the PSP to fine-tune the process for full capability to deliver customer requirements.
8. Evaluate results: this is to answer the following two questions:
   - Did the process work?
   - Do the results of what we did meet customer requirements?
9. Recycle: this is for continuous monitoring and the changing of steps following variations in customer requirement and exploiting best practice and new learning opportunities.

1.3 Driving quality strategically: the deployment process

Rank Xerox deploys its quality and benchmarking programmes using *hoshin kanri* (quality policy deployment). This process, illustrated in Figure 1.4 drives the vision through the establishment of goals and targets, their effective communication at all levels and putting in place mechanisms for monitoring, reviewing and controlling performance on a regular basis, and at all levels. Policy deployment has been defined as follows:

> Policy deployment interactively integrates a company's goals and related objectives at each organizational level, and then directs the application of resources at each level to the practice of business processes capable of meeting and exceeding the objectives and goals... Goal/objective congruence and focus is a key aspect and benefit of policy deployment. The system provides for detailed specification of the performance requirements for each business organization or level through a participative process of objective negotiation.

The final output from the policy deployment process is a document with roles, responsibilities, and objectives (RROs) for each employee highlighting their individual commitment to contribute to the company's goals. Policy deployment therefore is the process by which Rank Xerox translates its values and goals into a series of activities capable of delivering the desired outcomes.

1.4 Benchmarking at Rank Xerox: the beginning of a new phase

Benchmarking was introduced in the early 1980s as one of the new tools to assist Rank Xerox to restore its competitive advantage and eliminate complacency. David Kearns, former CEO at Xerox and the champion for the introduction of competitive benchmarking, explains the rationale behind the development of this powerful competitive tool:

> We took competitive analysis one step further and came up with what we now call competitive benchmarking. It's an intense, in-depth study of what we think is our best competition. It's a continuing, never-ending process, and it's an integral part of our new and stronger emphasis on quality. Every department at Xerox should be benchmarking itself against its counterpart department at the best companies we compete with. We look at how they make a product... How much it costs them to make it... How they distribute it, market it, sell it and support it... How their organization works... What kind of technology they have. Then, we all go back and figure out what it takes to be better than they are in each of those areas.

Xerox defines benchmarking as: 'A continuous, systematic process of evaluating companies recognized as industry leaders, to determine business and

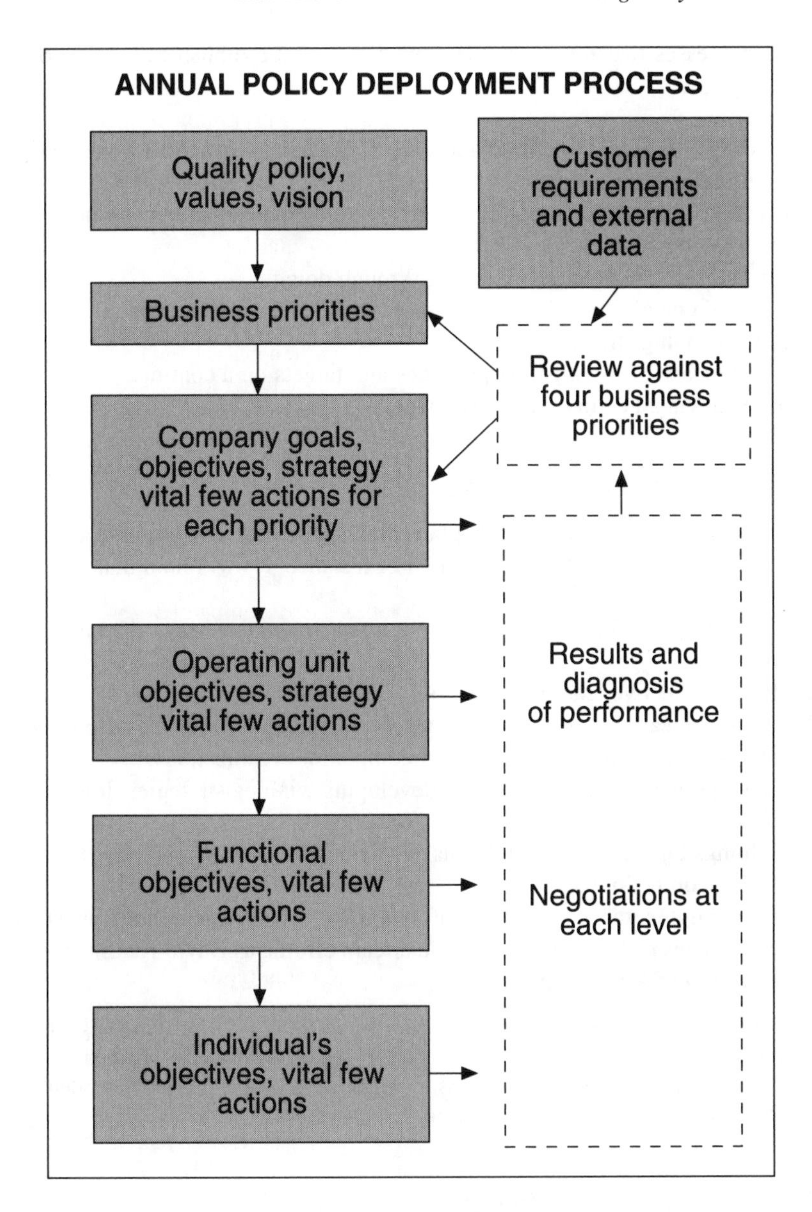

Figure 1.4 *The quality policy deployment process at Rank Xerox*

work processes that represent "best practices" and establish rational performance goals.'

- It is described as a structured approach for looking *outside* the organization for creative ideas.
- It is a learning experience for projecting performance targets and developing winning strategies.
- It is a discovery process – learning through doing.
- It is not a mechanism for determining headcounts or reducing staff.
- It is a winning strategy.
- It is a means of establishing priorities and targets on a continual basis, and for achieving a competitive advantage.

1.4.1 Benefits of benchmarking

At Rank Xerox it is recognized that benchmarking can lead to tangible and soft benefits in all aspects of business life. For instance, it was found that:

1. Benchmarking brings about newness and innovative ways of managing operations.
2. It is an effective team building tool.
3. It has increased general awareness of costs and performance of products and services in relation to those of competitor organizations.
4. It is a powerful methodology for developing winning strategies. It is a precise way of measuring gaps in performance.
5. It brings together all the divisions and helps to develop a common front for facing competition.
6. It highlights the importance of employee involvement and, as such, encourages recognition of individual/team efforts, as David Kearns, previous CEO of Xerox explains:

> Competitive benchmarking tells us where we have to go. It establishes goals which bring us productivity levels equal to or better than the competition. Employee involvement is the process by which we get there. It is problem identification followed by problem solving.

1.5 A sound methodology

The Xerox benchmarking processes use five stages and ten steps as illustrated in Figure 1.5.

Table 1.2 illustrates the purpose of each phase and the related steps.

Figure 1.6 illustrates the benchmarking template which integrates the internal and external necessary activities and links outcomes to means. It also high-

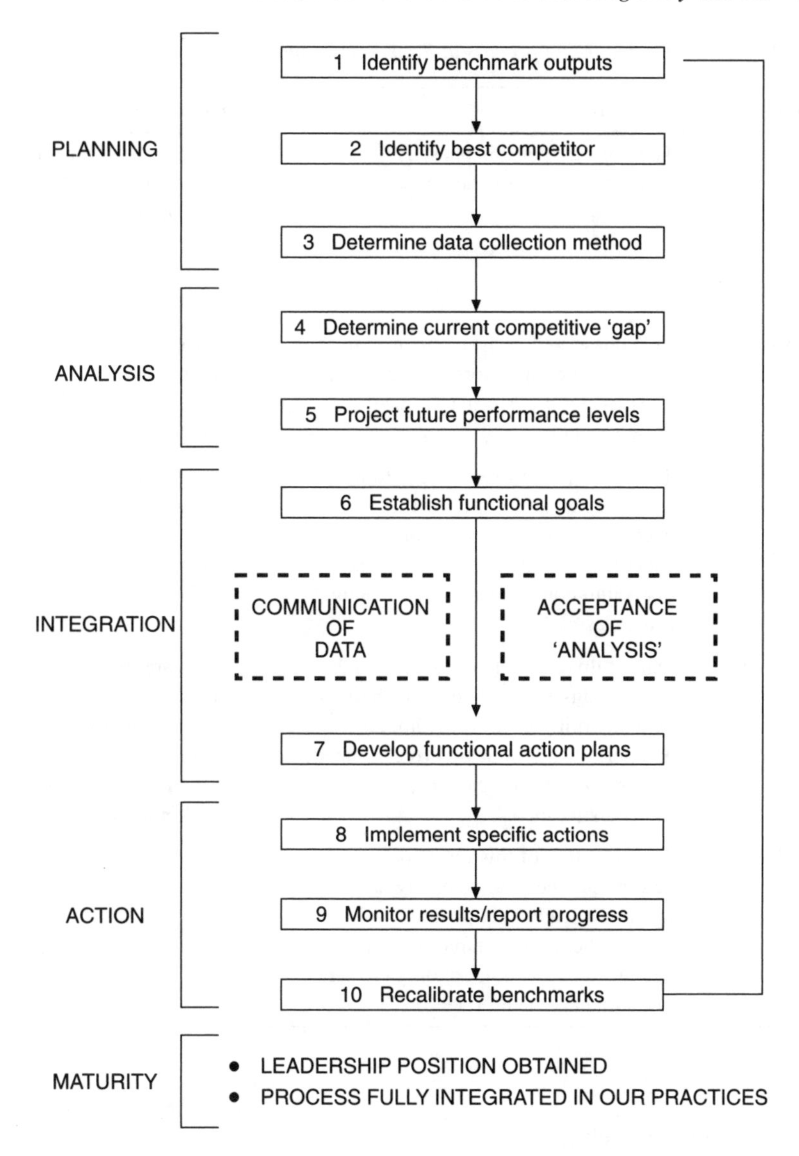

Figure 1.5 *The Rank Xerox benchmarking process*

lights the specific steps for each of the four quadrants concerned. The maturity phase is not included in the template since it reflects a review of the outcomes from taking action as a result of the benchmarking effort. The common pitfalls are listed in Table 1.3.

Table 1.2 *The benchmarking process at Rank Xerox: application aspects*

Phase	*Objectives*	*Relevant questions*
1 Planning	The objective of this phase is to prepare a plan for benchmarking	(i) What is the subject to be benchmarked? (ii) Who are the best competitors? (iii) What is the best data-collection method?
2 Analysis	This phase will assist companies to understand competitors' strengths and to assess their performance against these strengths	(iv) What is the current competitive gap? (v) What is the projected competitive gap?
3 Integration	The objective of this phase is to use the data gathered to define the goals necessary to gain or maintain superiority and to incorporate these goals into companies' formal planning processes	(vi) How are the results of the analysis communicated? (vii) What are the new goals?
4 Action	During this stage, the strategies and action plans established through the benchmarking process are implemented and periodically assessed (recalibrated) with reports of companies' progress in achieving them	(viii) What are the action plans? (ix) Is the company achieving its plan? (x) What is the plan for recalibration?
5 Maturity	The objective of this phase is to determine when the company has attained a leadership position and to assess whether benchmarking has become an essential, ongoing element of its management process	

1.5.1 Useful guidelines

In addition to a comprehensive list of points and issues that the benchmarking teams need to address in order to ensure an effective and successful outcome from the benchmarking initiative, Xerox has issued some specific guidelines for the external activities which involve benchmarking partners:

1 *Preparing for a telephone survey*

- Establish specific questions that need to be asked with a lot of clarity

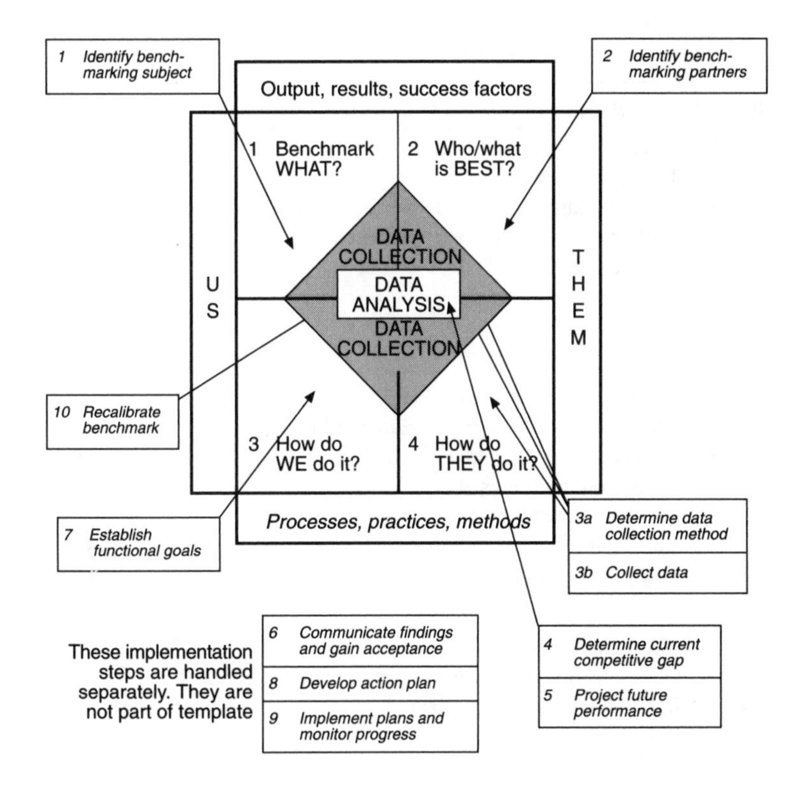

Figure 1.6 *The benchmarking template at Rank Xerox*

- Assemble information in similar areas which can be shared with partners
- Ensure that the contact(s) for the partner organization are the right people who can provide the right information
- Be very clear about purpose, objectives, requirements, expectations and mechanics of benchmarking initiatives
- Perhaps send a copy of the questionnaire to contact(s) to give them time to think about answers and assemble data, prior to the telephone survey taking place
- Conduct the survey and decide on whether there has to be a follow-up visit or another discussion on the telephone.

2 *Preparing for a company visit*. Most of the issues to be observed for a telephone survey can still apply in the context of an industrial visit. Among other things that need to be taken into account are:

- Issues of performance measures
- Methods of practices used

Table 1.3 *Effective implementation: what are the common pitfalls?*

Step no.	*Common pitfalls*
1	• Subject selected not critical • Too many subjects • QCDs (quality, cost and delivery) not considered • Customer satisfaction not considered • Too many metrics • Poor metrics • No purpose defined • No management buy-in
2	• Industry leaders not considered • Other divisions of own company not included • Customers not consulted • Public sources not used • No management buy-in • Companies in other industries not considered • Companies eliminated without adequate information
3	• Questions not related to subject • 'So what?' questions • Questions difficult to answer for your own organization • Interrelationships not probed • Practices not investigated • Public sources not used • Only one person making company visits • Key findings trusted to memory
4	• Analysis paralysis • Overprecision • Reasons for gaps not identified • Interrelationships not understood • Cause and effect not understood • Not keeping to subject • Value of practices not determined
5	• Overprecision • Step not completed
6 and 7	• Results of analysis poorly communicated and management/customer acceptance not obtained • No changes to goals – Benchmarking not integrated into the planning cycle – Management not committed to achieving benchmark
8–10	• Action plan not developed or approved • Action plan not communicated • Post-implementation not monitored

- Methods of calculating the performance
- Data availability and the issues of confidential and sensitive information.

1.6 Examples of applications of benchmarking at Rank Xerox

1.6.1 Distribution benchmarking study

Purpose of the study

Following a rationalization exercise, this benchmarking study was sponsored to examine leading organizations in logistics and distribution with the view:

1. To gain an understanding of other company practices in the following areas:
   - Organizational concepts
   - Physical operation
   - Warehouse facilities
   - Customer requirements
   - Order management
   - Financial management
   - Inventory management.
2. To establish a continuing dialogue with the companies to enable specialists from specific disciplines to visit their counterparts and discuss problems and procedures in greater depth.
3. If possible, to establish a management forum for regular (perhaps 6-monthly) meetings to debate future plans and strategies. This objective has been realized by the round-table conference set up by TSD in conjunction with the PRTM Business Consultancy.

Partners selected

Five partners were selected for this exercise:

- 3M EUROSPA (European spares), Dusseldorf
- Ford Motor Company, Car Park Division, Centre Cologne
- Sainsbury's Regional Depot, Buntingford, Hertfordshire
- Volvo Parts Distribution Centre, Gothenburg
- IBM International Spares Warehouse and French National Machines Warehouse.

Key learning

- The first fill order rate achieved elsewhere was between 93 per cent and 99 per cent while Rank Xerox achieved approximately 82–83 per cent.
- The warehouse procedures employed elsewhere gave productivity levels of between two and three times that achieved at Rank Xerox.

General recommendations were to work across functions covering the whole span of logistics. Specifically, this project led to the recommendations in Table 1.4.

Table 1.4

Area	*Recommendations*
Stocking echelons	• A team to be set up in order to implement inventory control routines based on engineer utilization
Systems	• To develop a single or several compatible databases with complete real-time access. This is to facilitate rapid response to the field thus giving engineers the confidence they need to reduce local inventories
Distribution costs	• Rank Xerox management should calculate precise logistics costs from central planning through the operating company distribution network in order to support the field inventory rationalization projects
Transport	• Rank Xerox to establish transport planning for distribution, raw materials, manufacturing components and sales as a separate strategic department
Warehouse developments	• Detailed investigation of pick/pack by customer or multiple customers for small and emergency orders using especially adapted trollies • Possibility of using picking trollies as the shipping containers in order to remove post-picking activities
Logistics developments	• Actions to improve first pick rate and to establish implications on inventory levels, customer satisfaction and procedures • Capitalization of spares (including diagnostic kits) • Repair strategy • Spare part identification • Handling emergency orders

Framework for analysis and benchmarking

A comprehensive document was put together to cover all the aspects that the benchmarking team would wish to examine:

1 Organizational aspects
2 Physical operation from receipt to despatch
3 Warehouse facilities
4 Customer requirements
5 Other management
6 Financial management
7 Inventory management.

Appendix 1.1 contains the detailed structure of the five visits undertaken and all the aspects examined by the project team and Appendix 1.2 illustrates the format for managing benchmarking projects. This is used at Rank Xerox for several purposes:

- To inject discipline into how projects are being conducted
- To monitor progress for each project
- To guide the teams concerned with key deliverables
- To ensure that there is an action plan at the end of the visits
- To build a total and full history for each project
- To form a database on areas of knowledge and expertise
- To provide consistency and a systematized manner for developing a culture of benchmarking.

1.7 Benefits and achievements from benchmarking within Rank Xerox

It will perhaps be difficult to quantify in specific terms all the benefits that have accrued over the years within Rank Xerox as a result of using the art of benchmarking. Nonetheless, since Xerox recognizes that benchmarking is the key driver for their superior performance, perhaps the ultimate benefits are those related to market position. Through the discipline of challenging its own practices, injecting new learning, stretching people and processes for high standards of performance, Rank Xerox not only achieved its desired status in world markets but also won recognition through gaining prestigious prizes and awards of quality excellence.

1.7.1 Xerox Corporation Business Products and Systems (BP&S)

When Xerox BP&S won the Malcolm Baldrige National Quality Award (MBNQA) in 1989 it recognized that the benchmarking process it developed and used was the key trigger for its major achievements which at the time included, among other things:

- Company-wide performance measurement covering around 240 key areas of product, service, and business performance
- Targets set are those of world leaders in the specific areas concerned, regardless of their industry sectors
- Significant returns from the company's strategy for continuous quality improvement
- Gains in quality including
  - A 78 per cent reduction in the number of defects per 100 machines
  - Greater increased product reliability as measured by a 40 per cent decrease in unscheduled maintenance
  - Increasing copy quality leading to a position of world leader
  - A 27 per cent drop in service response time
  - Significant reductions in labour and material overheads
  - The first company to offer a three-year product warranty.

1.7.2 Rank Xerox Ltd and the EQA Award

When Rank Xerox Ltd won the European Quality Award (EQA) in 1992 (and they were the first company to have won this prestigious prize) they achieved perhaps a global dominance as far as pioneering with quality was concerned. Like their experience in the USA following the winning of the MBNQA Award, Rank Xerox Ltd became a source of inspiration to thousands of European-based businesses. The winning of the EQA award did perhaps demonstrate that success with quality is not due to chance and can be replicated in different parts of the world, and also that quality knowledge and principles can be transferred. As in the USA, Rank Xerox Ltd used the same benchmarking methodology to support the leadership through quality strategy and to reinforce and strengthen various critical areas of the business.

Benchmarking has been used in different ways and at various levels by Rank Xerox Ltd:

1. To help develop the business planning process for the next 3–5 years by establishing goals and targets at benchmark levels.
2. Use by various individual operating units as part of their plan to close gaps in performance and achieve corporate goals and targets.
3. Use at corporate level to benchmark various European activities in support of the common goals.

4 An example of a benchmarking application is in commodity purchasing. The project was to improve ways of designing, procuring and manufacturing sheet metal products. This project led to new learning and major benefits, in the following areas:
   - Commonality in materials – through a reduction from seven types and thirteen thicknesses in 1986 to two types and eight thicknesses by 1989 with an 11.5 per cent cost reduction
   - A review of design practices to reduce processing steps, and an improved approach to acquiring tooling.

Other examples of benchmarking applications include:

- Warehousing/distribution
- Financial results, investments and market share
- Query and complaint handling
- Distribution in logistics
- Management development/training.

1.8 Business Excellence Certification (BEC): the way to effective integration of benchmarking

The BEC process was developed in 1991 as a method for continuous self-appraisal of Rank Xerox Ltd's overall quality performance. The elements of total business performance (Figure 1.7) measured by BEC include:

1 Management leadership
2 Human resource management
3 Process management
4 Customer focus
5 Quality support and tools
6 Business priorities/results.

The six key categories are then broken down into forty sub-elements with their own specific measurement targets. The self-assessment is validated by representatives from sister divisions and to obtain certification, units need to demonstrate:

- An operational command of leadership through quality tools and processes
- Good business results through continuous improvement
- Good prevention-based processes.

Guidelines for the assessment processes are illustrated in Figure 1.8.

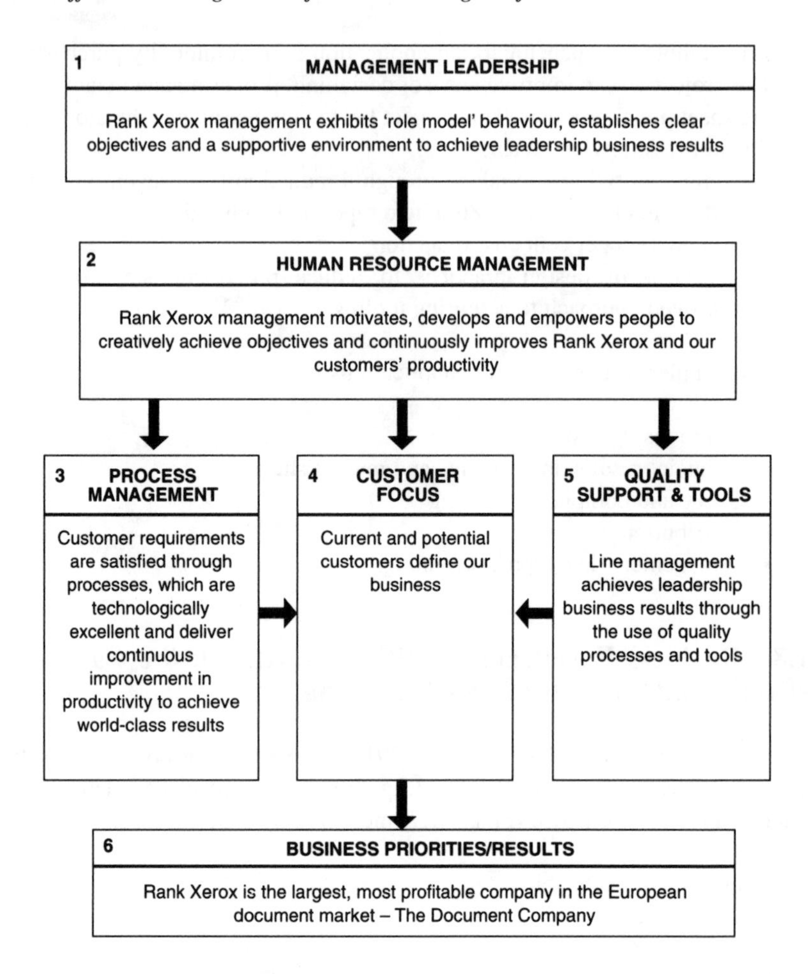

Figure 1.7 *The Business Excellence Certification framework*

1.8.1 Integrating with policy deployment

BEC challenges the way goals are developed and communicated, as a separate item on the BEC criteria. In addition, BEC assesses the effectiveness of Rank Xerox Ltd in achieving its goals (Figure 1.8). The process determines:

- Reports on quality achievements (successes and failures) + causes
- Progress with targets (four key targets as illustrated in Figure 1.8)
- The balance between enablers (categories 1–5) and results (category 6).

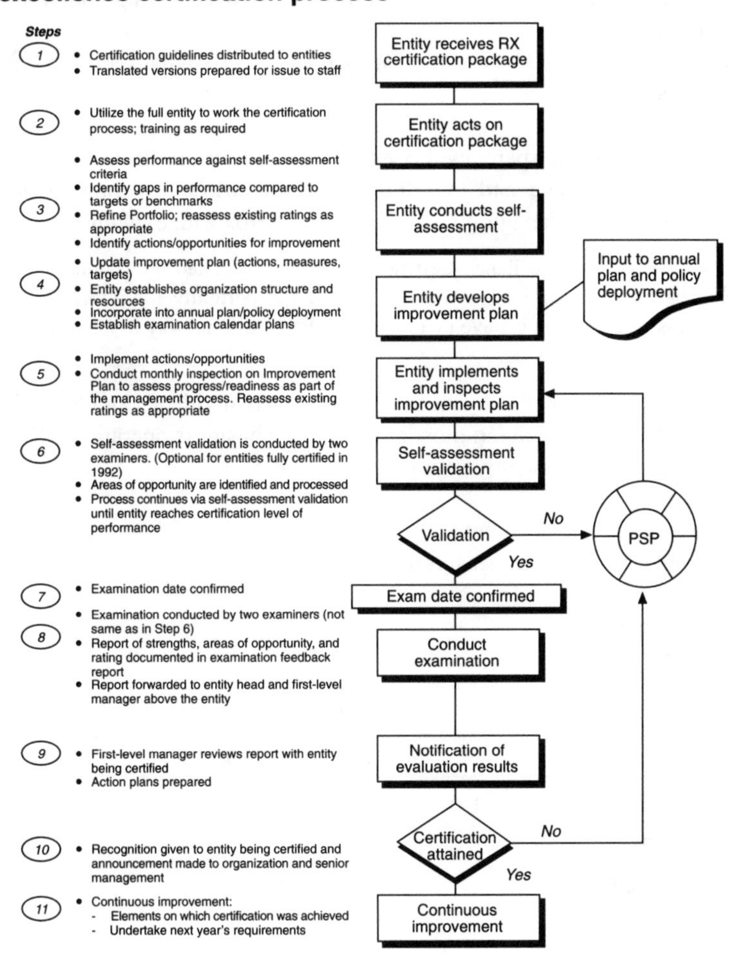

Figure 1.8 *Guidelines for assessment with BEC at Rank Xerox*

1.8.2 BEC and continuous improvement

BEC is a continuous process of improving in all areas of the business so that targets are met, benchmarks are set and the business is moving from strength to strength. Through ongoing monitoring and regular review of quality deployment, BEC can help in:

- Establishing root causes of success and failure
- Highlighting facilitating factors or obstacles in achieving the following year's targets
- Defining key actions to tackle identified obstacles so that set targets can be achieved.

BEC is based on transparency, honesty and the desire to find out about true levels of performance and therefore learn new ways for doing better. As Bernard Fournier (Managing Director), Rank Xerox Ltd, explains:

> 'I want Business Excellence Certification to be conducted in an honest, open way based on trust. This is a major learning opportunity for improving the business and we must behave in a mature and professional way to maximize the benefits of the experience.'

In 1993, BEC in its third year has started to pay dividends and Rank Xerox Ltd are reporting benefits in most areas of their business operations. Some of the areas which consistently scored high include:

- Cost/expense management
- Customer satisfaction measurement systems
- Management by fact
- Policy deployment.

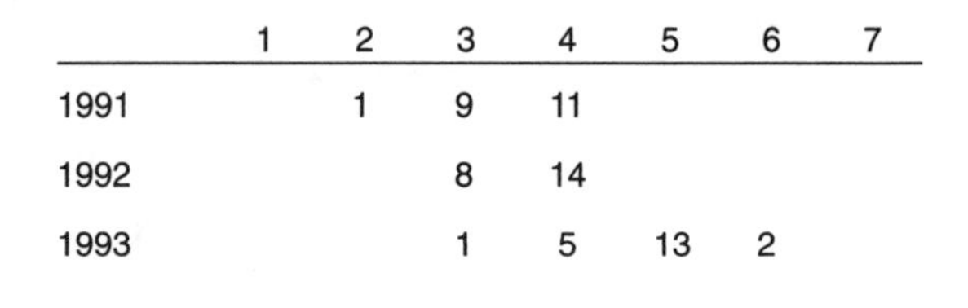

	1	2	3	4	5	6	7
1991		1	9	11			
1992			8	14			
1993			1	5	13	2	

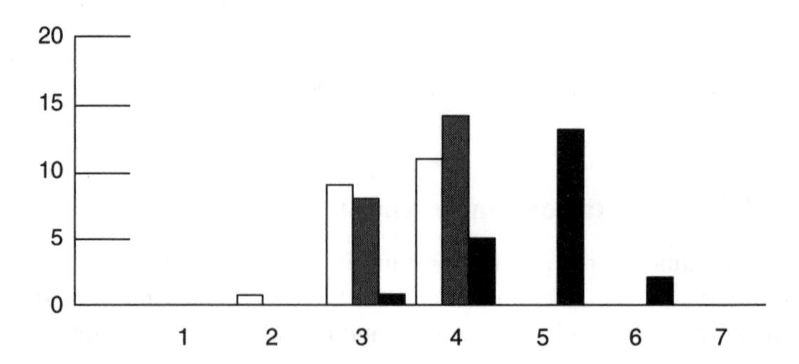

Figure 1.9 *Benchmarking pervasiveness at Rank Xerox*

Under the heading of support tools and specifically on benchmarking, BEC has helped indicate that benchmarking is increasingly pervasive and that a corporate-wide culture of using benchmarking is almost established (Figures 1.9 and 1.10).

1.9 Future issues

The journey of quality for Rank Xerox is set to continue and to move from strength to strength. In order to prevent levels of complacency in all areas, Rank Xerox decided to benchmark its benchmarking process in order to check its robustness, completeness, pervasiveness and overall effectiveness. This exercise was facilitated by the author who played the role of facilitating access to partner companies and to give stimulus and objective views on the approach. The author also assisted in developing the methodology used for the exercise (Appendix 1.3).

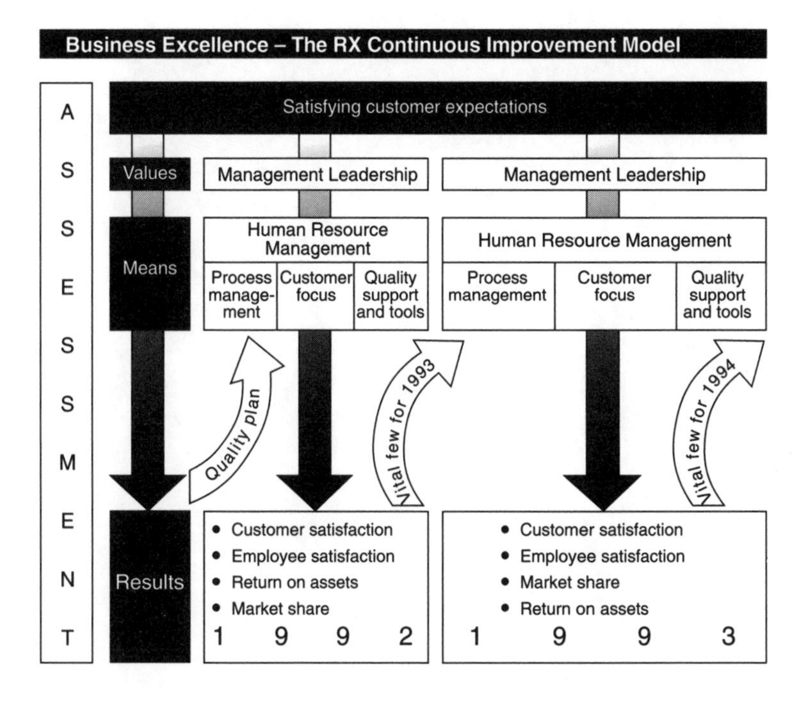

Figure 1.10 *The integration of BEC with quality policy deployment at Rank Xerox*

Acknowledgements

The author is greatly indebted to the following Rank Xerox staff for information and support given for writing this case study:

Paul Leonard, Senior Consultant, Rank Xerox Quality Services Ltd
Marion Walker, Rank Xerox Ltd
Sam M. Malone Jr, Project Manager, Xerox Corporation
Richard Cross, Senior Consultant Rank Xerox Quality Services Ltd

Appendix 1.1: Rank Xerox Distribution Benchmarking Project

Visit brief

The aim of this document is to assist in gaining an understanding of other companies practises in the following areas:

1 Organizational concepts
2 Physical operation from receipt to despatch
3 Warehouse facilities
4 Customer requirements
5 Order management
6 Financial management
7 Inventory management.

In addition to reviewing current practises Rank Xerox should look at how other organizations intend to change and improve in the future.

Section 1 Company profile

Please supply a brief profile of your company including:

- Which business sector it occupies
- Geographic spread of activity
- Market share
- Distribution network
- Turnover
- Overall management/organization chart covering the company's activities
- What is the role of distribution within the organization? Is it located at one end as a highly important integral part of the business or is it at the other end a necessary evil?

Section 2 Organization

- Please supply a detailed organization chart of the total distribution process from materials planning through warehouse management to the field engineering set-up.

Section 3 Materials planning

- Who estimates demand patterns?
- How is this done, manually or by computer?
- How far in advance must demand be committed to the suppliers?

- Are there different policies for established and new products?
- Do you have an asset recovery activity? If so, is this controlled centrally or locally?
- How do you take refurbishing activity into account when forward planning material?
- Do you have returns to the centre? If so, how are they controlled?
- How do you control excess/surplus?
- What is the write-off policy? What target percentage purchase?
- What is the actual percentage write-off?
- Do you have a product end-of-life policy?
- What are the configuration change rates?
- How do you handle seasonality?
- How do you control your internal manufacturing order book?

Section 4 Stocking policy/inventory control

- How many echelons do you have?
- Are there defined stocking policies for each echelon?
- What decided the number of echelons?
- What level of inventory is held in each echelon by category where applicable?
- How do you conduct cycle count activities?
- What computer systems do you utilize to assist in the inventory control activity and how are they linked to the materials planning systems?
- How do you measure inventory performance
  MOS – Forward or historical, or
  DOS – How do you report/calculate net inventory?
- For slow-moving parts how do you determine the break for local stocking versus rush ordering?
- Van kit
  - How do you determine van kit contents?
  - Lead time from echelon above
  - Order types
  - How do you control return of unserviceables and excess?
  - How do you measure service level?
- Lowest non-purchase echelon:
  - What parts and how do you determine?
  - Lead time/order types
  - How do you set up reorder points?
    Usage or demand
    Establishing for new product.

Section 5 Order control

- Is the organization reactive or does it operate against planned order levels?
- What order categories do you have and what are the volumes against these categories for average, peak and trough workloads?
- Is the activity seasonal?
- How many customers are serviced and where geographically does the bulk of the business come from?
- Are deliveries to field engineers push or pull?
- Do you supply to the piece part or sub-assembly level or both?
- Is order control handled via a computer system and how do you track it?
- How do you conduct your customer liaison/customer service business?
- Do you despatch orders to a prespecified shipping date or do you despatch on request?
- If the former, what is your performance against plan?
- Do you control the amount of time taken from order allocation to extract?
- What is the EO or rush order system?

Section 6 Warehouse facilities

- What storage system do you use (random/fixed)?
- What storage media do you use (narrow-aisle racking, gravity racking, etc.)?
- What are the dimensions of the racking, height, breadth, etc.?
- How many locations of each storage type are there?
- Is the warehouse computer controlled? If not are you considering this for the future?
- Do you segregate fast and slow movers?
- How many receipt and despatch docks are there?
- How do you handle prepacking?
- In your receiving operation what is/are:
  - The count philosophy?
  - Frequency of receipts?
  - Dock to stock time?
  - The links into your accounts payable routines?
  - GIT control?
  - Dues out (back order) control?
  - Quality control?
  - Work standards/drivers, how are they devised and who monitors them?
  - Number of locations per spare?
- In your picking operation do you have
  - Prime back quantities?
  - FIFO?
  - Random or planned volumes?

  - Quality and quantity control?
  - Work standards/drivers by labour or order or both?
- Do you control the amount of time taken from order availability for picking to the actual pick time?
- Could you give an explanation of your marshalling, packing and consignment management process?
- Overall, within the warehouse, what are your computer controls and what use of bar code or other labelling techniques do you make?

Section 7 Transport

- Do you consolidate shipments? If so, how?
- What shipping volumes do you have?
- Do you have special transport for emergencies?
- What is the consignment shipment profile?
- Do you use agents?

Section 8 Finance/customs

- Do you have an import activity?
- How do you handle
  - Duty reclaim?
  - Letter of credit?
  - Hazardous goods?
  - Strategic goods?
  - Permits?
- Do you have or use a bonded warehouse?

Section 9 Working conditions

- What are the hours of work?
- Do you work shifts?
- What is the payment basis?
  - Flat salary?
  - Productivity payments?
  - Piece part payments?
- Do you have works councils?
- How do you motivate your field engineering staff?

Section 10 Measurement factors

- How many months of stock are held at each stocking echelon and can this be given by category?

- What are distribution costs as a percentage of revenue?
- Within your central warehouse how many orders per individual are processed per annum? If possible, this should be given for the net warehouse labour and for the total labour including all administration and security staff.
- What is the cost of processing an individual order?
- What are order turn-around times by order category?
- What are dock to stock times?
- What is the hit rate against the planned shipment days, if applicable?
- What is the first fill order rate or, conversely, how many orders received go into back order status?

Appendix 1.2: Rank Xerox Benchmarking Project: Management Handbook

Benchmarking Study Record

Title:..

Sponsor:...

Start Date:....................................... Completion Date:..................................

Benchmarking Study Start-up

Sponsor:...

Description of the purpose of the study and reason for selection: ....

....

....

....

....

....

....

Study Team Leader:...

Meeting dates (record role & attendance)

Team Members							
....							
....							
....							
....							
....							
....							
....							

PHASE ONE – PLANNING (STEP 1)

Measurements (Step 1 gl 3) selected:

....

....

....

Summary of purpose of study (Step 1 gl 4):

....

....

....

....

Step one review (Step 1 gl 5)
Reviewed with: ....

....

....

On (date): ....
Outcome: ....

PHASE ONE – PLANNING (STEPS 2 & 3)

Benchmarking partners: (Step 2 gl 6)

....

....

....

Reasons for selection ....

DATA COLLECTION CHECK LIST

Process flowcharted (copy attached)	Y/N
Questions prepared (gl 1) (copy attached)	Y/N
Questions answered (gl 2) (copy attached)	Y/N

N.B. Must be completed prior to contacting benchmarking partner(s)

Data collection method(s) used (gl 4–6) ....

....

PHASE 2 – ANALYSIS

The benchmark is (Step 4 gl 1–3):

....

....

....

The current competitive gap is (gl 4): ....

....

....

....

Reasons (gl 5):...

....

....

Possible benchmark drivers (gl 6): ....

....

....

Projected competitive gap (Step 5 gl 1):

....

....

....

....

NB All supporting documentation must be kept filed with this record.

PHASE 3 – INTEGRATION

Results of analysis communicated (Step 6 gl 1–4)

To whom?	....
By what means?	....
By whom?	....
Outcome	....

New goals (Step 7 gl 1&2)

Current goals	Recommended new goals
....	....
....	....
....	....
....	....

Revised gap projection (gl 3) ....

....

....

Reviewed with (gl 4) ....

....

....

On (date) ....

Outcome ....

PHASE 4 – ACTION

Action plan developed (Step 8 gl 1&2) date ....

Reviewed with (gl 3)..

....

On (date)...

Outcome...

Action plan implemented (Step 9 gl 1) date...

Post implementation monitoring (gl 2)

Commenced date...

Key responsibility ....

Recalibration plan documented (Step 10 gl 1&2)

When ....By whom...

NB All supporting documentation must be filed with this record.

SUMMARY

A concise narrative of the success/improvement areas as a direct result of the benchmarking study:

....

....

....

....

....

....

....

....

....

....

....

....

....

....

....

....

....

....

....

....

....

....

Written by ....on (date)..

Storyboard	Measure of Output, Result of Success Factor
Team leader ______ Subject ______ Team members ______ Benchmarking partner(s) ______	
Our current process, method or practice	The BENCHMARK process, method or practice
Our PLANNED change	Target/actual change in our RESULTS

Appendix 1.3: Data collection tool for Rank Xerox's Benchmarking of Benchmarking Process Project

(A) Process

- How mature is your process?
- How many elements are there in the process?
- How was policy deployment introduced phased/timescales?
- Why did you support policy deployment, i.e. with what other processes, if any?

Objective setting

- How is strategic planning achieved at British Airways, e.g. at a workshop/by the 'top' team?
- How are targets set, e.g. uplift on last year's performance/voice of process/voice of customer?
- What is the relationship between goals and critical processes; is there an explicit test of congruency of objective with capability?
- How is ownership of a goal arrived at?

(B) Communication strategy

- What terminology do you use for communications, e.g. deployment/goal translation?
- How do you communicate the goals, e.g. conferences/meetings/newsletter?
- How do you get feedback/measure of effectiveness of the process?
- How do you ensure ownership, i.e. commitment versus compliance?
- What is the timeframe for communication?
- How many layers are there in your organizational structure?
- What format do your collaterals/deliverables take?
- How are your collaterals produced/who produces them/who checks content?
- Where do you encounter obstacles in the communication process?
- What is the key driving force for your communication process?

(C) Performance

- How do you activate target setting at operational (activity/task) level?
- To what level is ownership/involvement pervasive?
- How is performance measured – how are measures set (by the individual/by group?)

- What is the level of uniformity of measures in the organization?
- What training is available?
- What review mechanisms are in place?
- What actions take place from reviews (e.g. adjustment of process at local level, adjustment of target at high level)?
- At what level is feedback loop as the actions take place, i.e. how far back does the review process travel/how 'closed' is the loop?

(D) Support mechanisms

- Have you got a formal appraisal system in place for performance, for personal development?
- Does development of employees derive from performance appraisal?
- Is there a reward and recognition system in place?
- Is there a linkage with policy deployment and performance planning and appraisal?

(E) Drivers for quality policy deployment

- What drives policy deployment in your organization?
- What external feedback measures are used for resetting/setting targets (voice of the customer, voice of the process)?
- What level in the organization are they used apart from goal setting?
- How do you monitor business performance in the marketplace?

(F) Closing questions

- How effective is your policy deployment approach?
- What benefits have you achieved from using it?
- Have you quantified the benefits?
- What scope for improvement is there for the policy deployment process?

2 The strategic application of benchmarking for best practice: a prioritization tool

The knowledge of the World is only to be acquired in the World, and not in a closet

Earl of Chesterfield

What we have to learn to do, we learn by doing

Aristotle

When quality is viewed as being the number of products that conform to specifications, a company is already behind the eight ball

Aguayo

2.1 Introduction

There is definitely a better awareness of the 'art of benchmarking' and its application is spreading to encapsulate various organizational contexts including non-profit-making sectors such as health care, the army, local government agencies, among others. Indeed, examples of benefits which may be derived from the use of benchmarking are in abundance and range from cost and time reductions, quality improvements, and new learning.

2.2 Benchmarking methodologies

Most texts, however, describe benchmarking as a 'tool' of quality. This is indeed both unfortunate and incorrect. One refers to the subject of benchmarking methodologies, and this is perhaps an area which is easy to describe as a tool. One could specifically refer to the Rank Xerox approach which is described in five phases and ten steps. The application of benchmarking, however, differs from the application of other tools and techniques of quality management. If one refers, for instance, to the use of statistical process control (SPC), most of its applications are at an operational level in order to provide local benefits.

Most tools of quality management are 'task related' and are used to reinforce the importance of data collection and the gathering of facts for decision making and problem solving. Benchmarking is perhaps much more encompassing as a concept, since its main focus is on larger processes which may not necessarily have local/operational impacts on the business but very much impinge on the level of competitiveness of the organization concerned.

Watson [1] in his book *Strategic Benchmarking* writes:

> What issues are addressed by strategic benchmarking as opposed to 'operational' benchmarking? Among the issues are: building core competencies that will help to sustain competitive advantage; targeting a specific shift in strategy, such as entering new markets or developing new products; developing a new line of business or making an acquisition; and creating an organization that is more capable of learning how to respond in an uncertain future because it has increased its acceptance of change.

It is evident from the above description that strategic benchmarking places more emphasis on knowledge and learning as the major source of competitive advantage, rather than on conventional means such as new technology, range of products and services. In other words, strategic benchmarking focuses on *soft* rather than *hard* aspects of competitiveness. The former are more likely to give *sustainability* and continuity of the momentum of progress and advancement.

The notion of benchmarking as a 'tool' is therefore incorrect since the impact of its application is more for changing attitudes and behaviours and raising commitment through better education, awareness and inspiration from model companies. Benchmarking is perhaps the best means for servicing the human asset by continuously supplying new ideas to sustain superior performance levels.

2.3 What is the scope?

It is very important to highlight that the strategic application of benchmarking has a lot of scope and certainly does not just relate to model organizations only. Benchmarking is relevant to any organization committed to the ethos of continuous improvement. It is a fallacy to think that the scope is only relevant to organizations wanting to become *number one* in their industry sectors. Depending on the learning curve, resources committed, and pace of achievements, benchmarking can lead to:

- Incremental improvements to existing performance standards
- Quantum leaps by instigating new practices and ways of working
- The road to excellence: creating the learning organization.

The above three objectives can be achieved through a strategy of closing a performance gap based on determining what the existing standard is (base line), what should really be the internal standard (entitlement) and developing an action plan for the immediate future, to achieve the entitlement. Benchmarking stretches organizations further by encouraging them to aspire for superior performance (desired goal) (Figure 2.1).

2.4 Benchmarking and strategic planning: is there compatibility between the two?

Strategic planning is the framework by which organizations can work with confidence and with a level of uncertainty and risk rendered to a minimum. Effective strategic planning makes organizations pass specific milestones to achieve various prizes and will ensure that the organizations concerned are 'on course' with the ultimate prize, that of achieving their desired mission. Strategic benchmarking, on the other hand, is the means by which their plan is determined. It is the process by which the vision/mission is established and challenging goals are developed. As Watson [1] argues:

> . . . This is the essence of strategic benchmarking and the link to a company's planning process. Companies selected for benchmarking because of their key

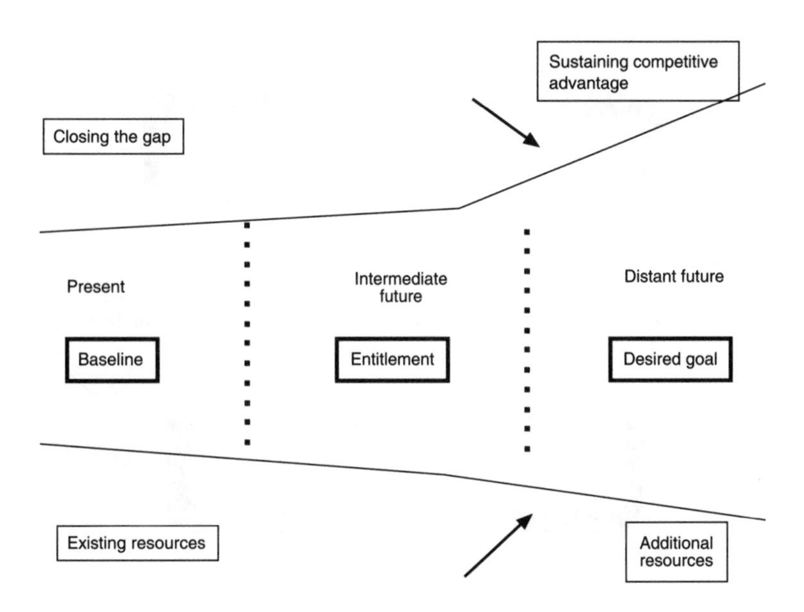

Figure 2.1 *Benchmarking encourages organizations to aspire to superior performance*

business process knowledge and performance indexes can serve as a basis for establishing challenging, yet realistic and achievable goals.

Conventional strategic planning tended to focus more on *what we need to happen* in isolation from internal operations. Many strategic plans fail to deliver because the targets are set too remotely from the processes which are expected to deliver and the basis on which targets are set is very often questionable, based heavily on 'individual ambition' rather than on 'wisdom and vision'.

Strategic benchmarking has added a new dimension by making strategic planning more effective and target setting a more systematic process based on reality rather than merely stretching organizations with unreasonable expectations.

2.4.1 Linking goal development to continuous improvement

Strategic benchmarking ensures that the process of goal development and deployment is dynamically managed through a closed-loop process (Figure 2.2). Through regular measurement and monitoring, action can ensue that

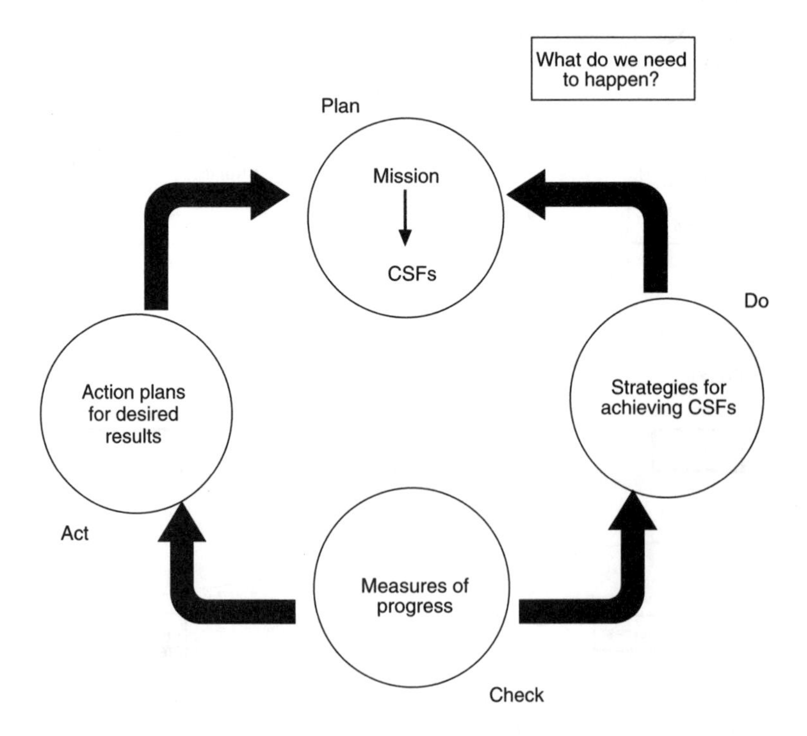

Figure 2.2 *The links between mission, CSFs, strategies, measures and action plans*

objectives can be achieved. The continuous cycle of improvement through plan–do–check–act is very applicable in this context. *This stage determines BENCHMARKS by focusing on various competitive parameters and performance standards of competitors and best in class organizations.*

2.4.2 Linking goal deployment to process management

This is perhaps the biggest contribution that the application of strategic benchmarking can make. Strategic plans are perhaps beautiful blueprints, but they need to be applied by asking the second most relevant question: *What do we need to manage, control and improve?* This stage of the application of strategic benchmarking determines the *capability* of the organization in achieving its desired vision and indeed in its success for superior competitiveness (Figure 2.3).

The external perspective of competition is perhaps very well known to most senior managers. Scanning markets, compiling benchmarks and studying competition very closely are activities which are very common. What is perhaps new is measuring the internal strengths, performance outcomes against other

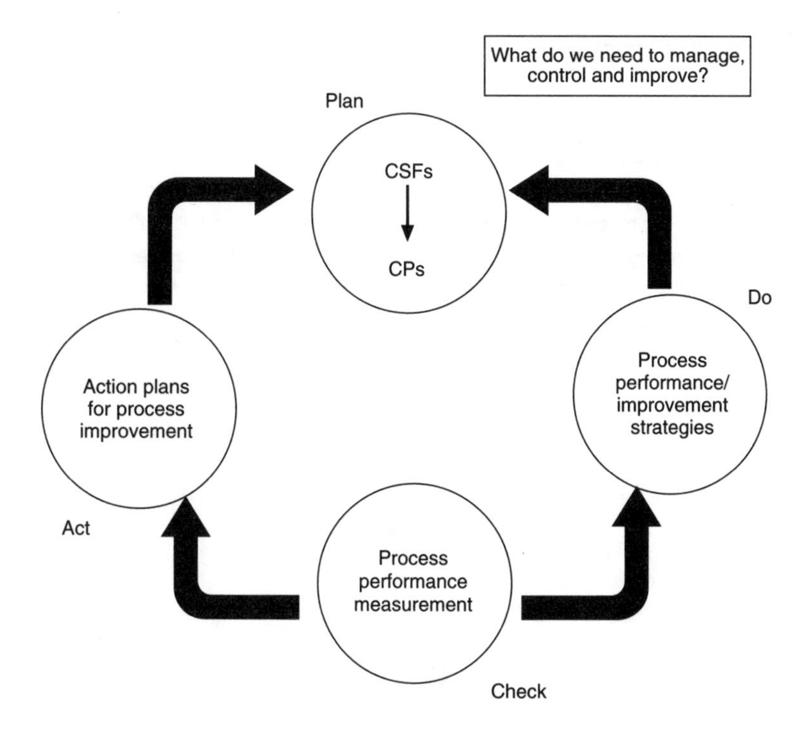

Figure 2.3 *The links between CSFs, CPs, strategies and measurement*

standards in order to develop stretch goals and achieve greater leaps in competitiveness. This stage does also ensure that:

- Goals are not deployed in isolation from the process
- *Capability* of the process is raised at the strategic level
- Performance measurement becomes a corporate-wide activity
- The focus is on the dynamics of the processes (*practices*) and not just the outcomes (*absolute measures*).

This stage determines priorities for BENCHMARKING and ensures that optimization takes place to provide uniqueness and competitive superiority.
Both stages are linked together by measurement activity on a continuous basis. As Figure 2.4 illustrates, regular measurement ensures that:

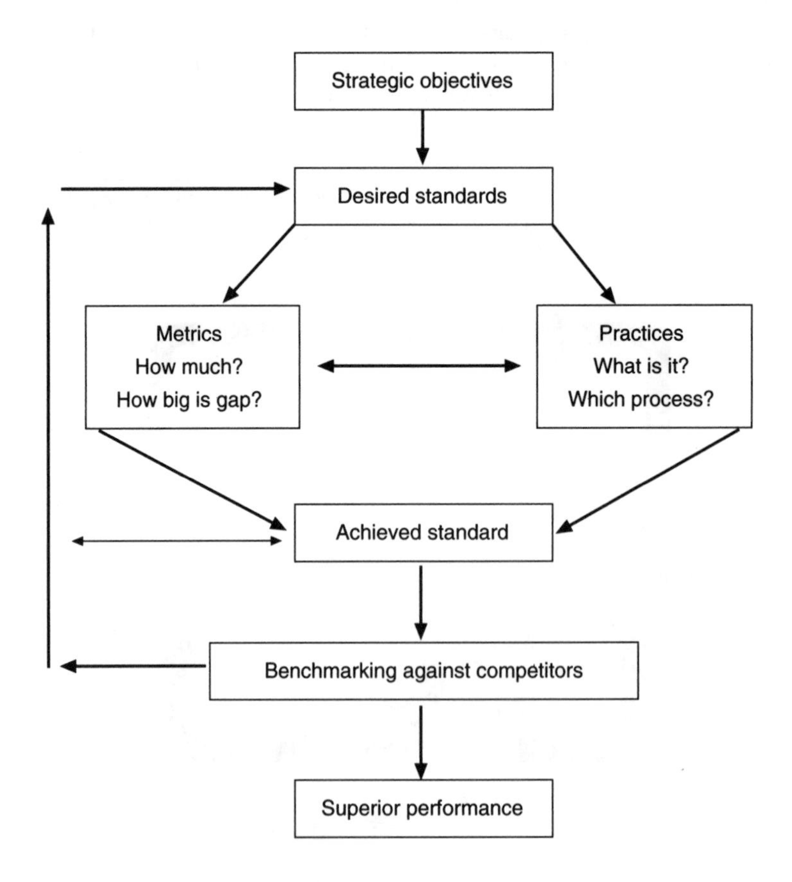

Figure 2.4 *Performance measurement*

1 Processes are improved to perform to the expected level.
2 Targets and objectives are achieved.

Continuous measurement is therefore important not just from the point of view of steering and taking corrective action but also for setting futuristic goals and stretching the organization concerned towards superior performance.

2.5 The implementation of strategic benchmarking: a prioritization methodology

Many questions are asked about the ideal number of benchmarking projects to be started, the areas where benchmarking activity needs to be focused and the establishment of priorities for benchmarking together with the resource implications for conducting every benchmarking project. In a sense all these questions are asked by organizations which embark on benchmarking activity in a very *ad hoc* manner, following opportunity rather than managing through discipline.

Some writers mention that perhaps as much as 85 per cent of benchmarking activity is done at an operational level possibly more. This is an indication of the spread of activity at various levels within the organization without effective coordination and certainly without perhaps asking the most basic questions:

- What do we want to benchmark and why?
- How relevant is the area to our strategic plan?
- How prepared are we to embark on such initiatives?
- What are the likely benefits we should expect?
- What is the likely impact expected on our competitiveness?

A prioritization methodology is therefore needed to address the above questions and many others. Prioritization ensures that the focus of benchmarking activity is on the core aspects of the business, the impact from benefits derived is closely linked to strategic intentions, and focusing on the 'vital few' resources will not become an issue.

The prioritization matrix for benchmarking is a tool which was developed at the University of Bradford [2] to assist organizations to choose core areas relevant to their business needs.

2.6 The application of strategic benchmarking within D2D: A case study

Design to Distribution Ltd (D2D) is a wholly owned subsidiary of the ICL group. It gained this status in 1993 having previously been ICL Manufacturing Division. Since 1991 it has strategically penetrated the contract manufacturing market and now aims to provide a total electronics manufacturing service to external customers. It has developed a total solution capacity whereby it is able to offer its customers' expertise, advanced manufacturing technology, and purchasing power to enable them to release their capital and resources to concentrate on their core competencies.

D2D have an established total quality culture and have benefited significantly from the resultant enhanced customer satisfaction and reduced cost of non-conformance. It constantly searches for process improvement tools and benchmarking is one tool D2D employs. However, D2D has concentrated benchmarking efforts on results-driven studies carried out by consultants. Although this provides a useful introduction to benchmarking, the next step for D2D to achieve its goal of world-wide competitiveness is a more process-driven approach. This study reviews the construction of benchmarking projects previously carried out by D2D and suggests a tool to help overcome one of the main pitfalls of process-driven benchmarking, that of appropriate process selection.

2.6.1 Importance of quality

As a contract manufacturer, it is imperative that D2D has a quality reputation in its marketplace. This means quality in every aspect of their business, in product, delivery and service. D2D operate in a rapidly growing market. This is because many electronic products are in the mature stage of their life-cycle, therefore producers will look to outsource production as it gives them more strategic discretion i.e. they are able to switch production on and off without having to open and close production facilities. For D2D operating in this market, total quality is essential for three main reasons.

First, many of the components D2D produces are used in very high value products. For example, they frequently produce Printed Circuit Boards (PCBs) for mainframe computers which make up only a minute fraction of their overall cost. If the PCB failed the loss of reputation and goodwill would be disastrous. The importance of defect-free production cannot be understated.

Second, the selling price of their products and services must be low enough to tip a customer's make or buy decision to buy. Yet they must still make a significant profit to maintain the R&D and production facilities necessary to maintain world-class manufacturing status. Right-first-time, to minimize the cost of non-conformance, is a concept all D2D employees have come to understand.

Finally, to be able to respond to a highly volatile marketplace, in terms of both technological advances and demand fluctuations, D2D must be aware of and able to meet the changing demands of its customers. D2D must be flexible, adaptable and responsive. Total quality with its unwavering focus on the customer provides D2D with the necessary market orientation required to proactively respond to customers' needs.

2.6.2 D2D: The European quality benchmark

D2D recognized the need for total quality and began a total quality initiative when still part of ICL in 1986. They have been extremely successful in establishing a total quality culture and found considerable benefit from the tools, systems and teams that quality management provides. This quality culture has been recognized externally with success in winning prestigious awards such as the UK best factory, a British Quality Award and a European Quality Award prize (1993) and have just received the European Quality Award itself. This makes D2D the first truly European company to win this coveted award.

All D2D's processes have been identified and documented and performance measures have been devised and implemented to ensure that continuous improvement is ingrained in the company culture. Employees recognize the need to focus on customers, both internal and external, and to meet and, if possible, exceed their requirements. This has been achieved by creating an equality and respect in the workplace by dramatically strengthening communication, by encouraging teamwork, by aligning the values and goals of the workforce with those of the organization, by giving the front-line workforce more authority to take decisions and by redefining the role of management from planning and control to leadership and support.

2.6.3 The introduction of benchmarking – database analysis

D2D are continually looking for further process improvement opportunities and benchmarking is one of many tools they have begun to employ. As has been well documented in the past, many organizations become overwhelmed with the task of benchmarking. Many are enticed by the evidence of the success of benchmarking from some of the world's leading organizations, and yet do not know how to even begin to benchmark effectively. Daunting questions such as What and who is the benchmark? How do we find it? Which processes do we benchmark? etc. are frequently raised. Also many organizations, with a hangover from competitive analysis, immediately consider that only competitors are worth benchmarking.

D2D began benchmarking by commissioning and participating in benchmarking studies carried out by various management consultants. Most frequently this entailed the utilization of database analysis to establish benchmarks and gain some qualitative best practice information. Selection of an area for study was largely based on a subjective assessment of the processes that were perceived to be performing the worst. No consideration was given to issues such as the impact a process has on customer satisfaction, employee motivation, business results or the ease with which the process can be effectively benchmarked. The ease of benchmarking is an especially important consideration in high-technology companies where innovation is the source of much competitive advantage. The process of innovation is widely regarded as one of the most difficult to benchmark.

The construction of benchmarking exercises was largely a team-based approach with an emphasis on cooperation between consultants and D2D managers. These teams have been used to both analyse 'as is' process performance and to implement benchmarking findings. The construction of benchmarking projects and teams is shown in Figure 2.5.

The early stages of benchmarking using consultants are very similar to that of an established benchmarking methodology (for example, the planning and analysis stages of the Xerox approach). Essentially the construction of benchmarking teams and analysis of present performance is methodologically very sound.

The database against which D2D's processes were compared is derived from numerous companies in various industries. Quantitative data in the form of process performance metrics and qualitative data in the form of questionnaires are collected from each organization. The consultants describe their role

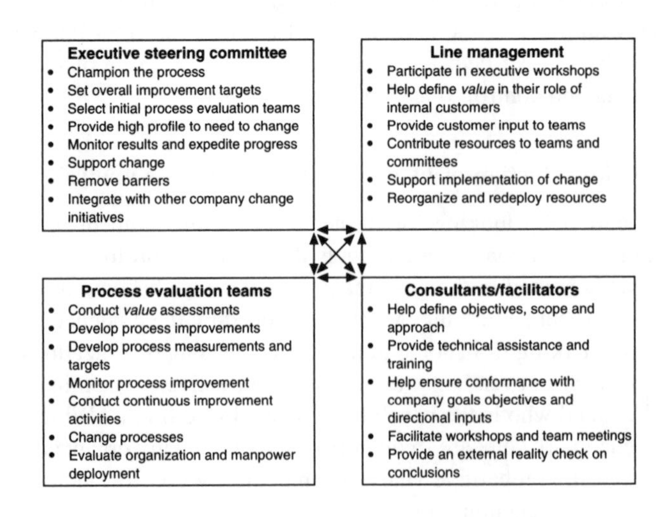

Figure 2.5 *Project organization – teams and roles*

as providing confidentiality and consistency in the interpretation of data from questionnaires during the analysis phase. Also, they look to gain universal acceptance for the objectives set, and in this respect, the opportunity to meet with benchmarking partner companies is suggested as a major contributor to commitment and success.

Early benchmarking studies carried out by D2D simply used the database analysis of the consultants to determine benchmarks and to set objectives. Although clearly this is not classical process-driven benchmarking, as an introduction to benchmarking it has made a very useful starting point.

It encouraged D2D to look outwards to gain an external perspective that even some TQ companies do not have. It also encouraged management by fact. One of the main benefits of this type of benchmarking, frequently quoted by D2D managers, was that it made them realize that they were not as good as they thought in certain areas. As a result, they were able to direct performance improvement efforts to those areas. In many cases this potential threat would not have been recognized by traditional financial or competitive analysis.

2.6.4 Limitations of database benchmarking

There are, however, limitations to what can be achieved through this type of analysis of benchmark data. D2D have recognized these limitations and have recently moved towards full process-driven benchmarking. The major limitation of database analysis is that to be able to make a real 'apples to apples' comparison between the performance data contained on a database and the performance of a process it must be absolutely certain that:

1 The process that is being operated is similar enough to those that make up the database to present a meaningful comparison, i.e. that it uses similar inputs, performs the same value-adding activities and provides similar outputs.
2 The performance measurement data are collected in exactly the same way. Percentages can well hide the many different ways of calculating performance measurement data.
3 The data in the database is kept current. Comparisons must be of what is currently achieved by others, not what they achieved in the past.

Another major limitation of this type of analysis is the fact that ownership of the process being benchmarked is perceived, especially by process owners, to be by a third party. Many D2D managers expressed reservations as to, first, how relevant the data were (for the above reasons) and second, how there could be motivation to change generated when process owners were not included in the benchmarking exercise. Many managers stressed that merely identifying performance gaps does not trigger improvement efforts. The organizations that have gained most benefit from benchmarking, e.g. Xerox, AT&T, Motorola, etc., all cite involvement of process owners as a critical factor for successful benchmarking.

In recognizing these limitations D2D have begun to consider a more process-based approach to benchmarking. Database benchmarking, although highly valuable for companies inexperienced in benchmarking, is still predominately a results-driven approach where the emphasis is on the ends and not the means, the whats and not the hows. The only indication of how a superior performance level is achieved is derived from the consultants' analysis of the qualitative data provided by benchmark companies. As a result, recommendations tend to be prescriptive, with the features of a best practice process being suggested for other companies, with little understanding of the appropriateness or the feasibility of the changes.

2.6.5 Process-driven benchmarking: the importance of goal deployment

D2D have stated a commitment to process-driven benchmarking. They realize that benchmarking is a way not only to study best-practice companies but also to adapt their ideas and creatively emulate their processes to dramatically improve performance in areas critical to their business success. They are committed to incorporating full process-driven benchmarking into their total quality programme to accelerate the cycle of continuous improvement.

If process-driven benchmarking is to be established as a company-wide quality-improvement tool, a carefully defined strategic direction is required that is understood by every member of the organization. Deployment is therefore vital, as the processes themselves and their role in driving the organization towards its mission must be understood. In D2D the goal deployment process is well developed and widely regarded as representing best practice. They have successfully aligned the employees of the business, their roles and responsibilities with the organization and its processes.

Manager and employee teams regularly review the performance of their processes and ensure that all processes have appropriate measurements, targets and benchmarks. As a result, TQ companies, such as D2D, represent the ideal environment for effective process-driven benchmarking to prosper.

D2D use the mission-critical success factors-critical process approach to goal deployment, with strategic and business reviews used to identify which processes are critical to the success of the organization. The current mission statement of D2D is:

> To become Europe's leading electronic contract manufacturing company, in our chosen markets by providing products and services world-wide that exceed our customers' requirements.

The critical success factors (CSFs) necessary to achieve the mission have also been identified and documented. They are listed in the EQA submission document as follows:

1. Total solution capability; to be able to provide a complete service for our electronics manufacturing customers
2. The best product quality
3. The best service quality
4. The best process quality as *benchmarked* against our competitors
5. Optimum organizational capability – to have skilled, well-trained people capable of responding to all customer requirements, and to changes in the business environment
6. The lowest time to market for new products
7. The best procurement capability – to work with our vendors to reduce our production costs, while working in long term partnerships
8. The lowest cost of ownership for our products – hence we must be the lowest-cost producer
9. The best relationships with our suppliers and customers, to be their preferred partner
10. The best technical capability as *benchmarked* against our competitors.

Importantly, every CSF has one measurable objective supported by many non-financial business results. This aims to provide a link between results and enablers. It also demonstrates D2D's commitment to performance measurement which has become a well-established tool to encourage process improvement.

From the CSFs a number of critical processes (CPs) have been identified. These are the processes that have to be carried out particularly well if the CSF, and therefore the mission, is to be achieved. The processes that are critical to the achievement of organizational goals are usually those where performance-improvement tools (such as benchmarking) are usually concentrated. The critical processes of D2D are;

1. System product and service delivery, including customer satisfaction and performance measurements for each of our six business streams
2. Prospective customer/strategic partner identification
3. Competitive technology status and *benchmark* identification
4. People satisfaction – investing in people processes
5. Process improvement reviews
6. Supplier partnership involvement
7. Cost reduction
8. Self-assessment
9. Recognition
10. Deployment.

Table 2.1 *D2D's critical processes and related success factors*

Critical process	*Related success factor*
System, product and service delivery, including customer satisfaction and performance measurement	1 Total solution capability 2 Product quality 3 Service quality 6 Time to market 9 Preferred partner
Prospective customer/supplier partner identification	5 Organizational capability
Competitive technology status and benchmark identification	5 Organizational capability 6 Time to market 8 Lowest cost of ownership
People satisfaction – investing in people process	5 Organizational capability
Process-improvement reviews	4 Process quality
Supplier partnership improvement	7 Procurement capability 8 Lowest cost of ownership
Cost reduction	8 Lowest cost of ownership
Self-assessment	10 Technical capability
Recognition	5 Organizational capability 9 Preferred partner
Deployment	All success factors

D2D have also assessed which CP applies to which CSF. In this way the process owners are able to envisage a direct link between their process and the CSF to which it relates. The interrelationship between the critical processes and the critical success factors is shown in Table 2.1.

It is important to mention the role of the European Foundation for Quality Management (EFQM) model, which is used for self-assessment. D2D have found it an extremely useful tool for them to establish their quality position and to develop quality improvement goals. The model ensures that the plan–do–check–act cycle of continuous employment is employed in each process throughout the organization. The three major businesses in D2D (Bare Board Manufacture, Printed Circuit Assembly and Purchasing and Materials Supply), all assess themselves against the model. D2D's interpretation of the model is shown in Figure 2.6.

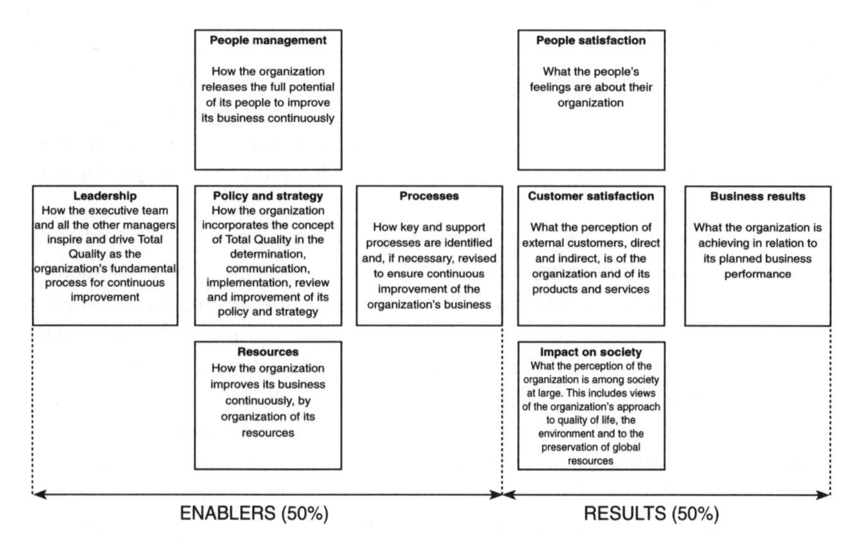

Figure 2.6 *D2D's interpretation of the EFQM self-assessment model*

2.6.6 Process-driven benchmarking: factors influencing process selection

Clearly, the critical processes are the major candidates for benchmarking, although this may extend to the sub-processes if the critical process itself is not at an appropriate scale or is too 'sensitive' to be benchmarked. This is frequently the case if the critical process is vital to the organization's competitive advantage.

These sub-processes may be a more suitable scale to be benchmarked, plus other organizations may be more willing to share information on their processes at this level. In the case of high-technology electronics manufacture there are many processes where willingness to share is a significant deterrent for competitive benchmarking. Consequently, the search for sub-processes that are more generic in nature becomes more pertinent as non-competitive benchmarking can then be contemplated.

The scale of the process for benchmarking is an important issue. If the study is too large it will consume many valuable resources and may produce large amounts of confusing data that are of insufficient detail to be used for process adjustment. Conversely, it is also extremely easy to fall into an 'activity trap' where small processes are analysed in such detail that the results of the exercise do not outweigh the benefit.

Essentially where benchmarking resources are limited, the decision as to which process to benchmark can be extremely difficult, even in a total quality organization where the processes themselves are well defined. Many tools and techniques have been suggested to enable an organization to reach a rational decision, and yet frequently a results-driven selection approach predominates. In these cases an organization simply looks at the performance of its processes and those that are doing the worst are benchmarked. No consideration is made of the potential impact that the processes have on the success (or survival) of the organization.

Organizations that have not introduced a good-quality system will find even the identification of processes difficult. These organizations will frequently turn to departments that they perceive are producing results that compare unfavourably with the competition. They will then undertake the well-documented activity of 'industrial tourism', which usually takes the form of a hands-in-pockets site visit followed by a large lunch. Not surprisingly, the results of this type of 'benchmarking' are largely unsatisfactory.

Many organizations will also immediately focus on the areas that they believe are the origin of most 'quality problems' – the shopfloor.* The factory is the area where visible and capital- or labour-intensive processes take place, therefore it may seem the obvious place to start benchmarking.

However, if an organization depends on vendors for 50 per cent or more of the cost of goods, then supply management is a more important function to benchmark than any factory-floor process. Superior execution of supply-chain processes such as order fulfilment, procurement, inventory management, etc. will have a far greater impact on profitability and competitive position than any production process.

Some processes also influence many more costs than they actually consume, and this is especially the case in high-technology manufacturing companies such as D2D. For example, the concept and systems design phases of technical product development typically consume 5–15 per cent of product life-cycle costs. Effective product developers understand, however, that these early phases determine approximately 85 per cent of life-cycle costs and even more of the pricing latitude available to the product. Therefore, these early design phases are the most critical influence over a product's ultimate financial success or failure.

Excellent execution of the early design phases is clearly essential, and benchmarking their performance is more beneficial than detailed examination of seemingly more costly later design stages. It also explains why many Japanese companies look to engineering competence and management systems (e.g. quality function deployment) to 'front-load' design activities to resolve manufacturing, sourcing and maintenance issues early in the design process.

Benchmarking priorities will vary from industry to industry and will, to some extent, depend on the amount of benchmarking experience an organiza-

*See Philip Crosby's comparison of this activity with 'where the police look for crime' (*Quality is Free*, 1987, p.17).

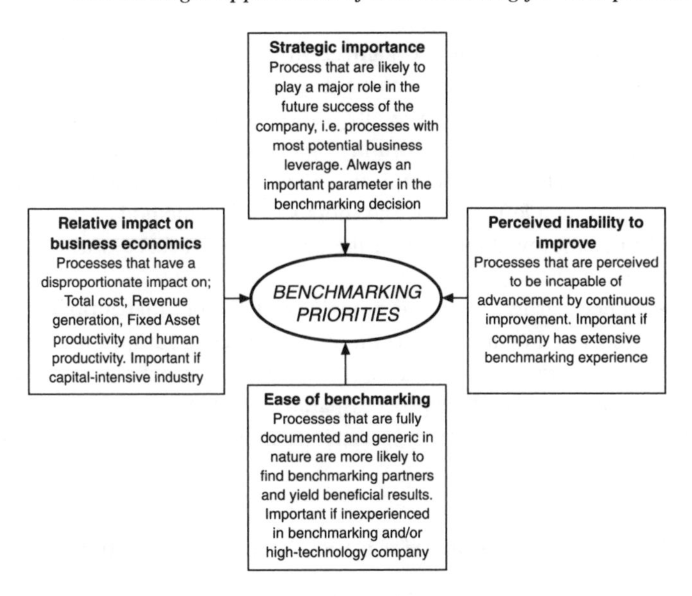

Figure 2.7 *Factors influencing the benchmarking decision*

tion has had. In general, the prioritization process must consider the following aspects of the organization's processes; future strategic importance (potential business leverage), economic importance, perceived inability to change and the ease with which the process can be benchmarked effectively (see Figure 2.7).

The vast majority of companies do not use any kind of systematic assessment of their processes before setting benchmarking priorities. These organizations frequently find that benchmarking significantly improves the performance of that process, but has little impact on overall efficiency, productivity or the bottom line. Prioritization is therefore an essential part of the benchmarking process itself.

2.6.7 Process-driven benchmarking: prioritization tools

The first step of the Xerox methodology is 'identify an area for study' and frequently this can be one of the most difficult (and arguably the most important) steps of the entire benchmarking process. Essentially benchmarking, like any other process, must be right-first-time for optimal use of limited resources. Correct process selection is therefore vital.

There are existing process prioritization methods available, such as the IBM Maturity Index and Xerox's use of Ishikawa diagrams. However, these selection methods are limited in terms of their capability to assess processes along different dimensions and most organization do not use them. Therefore, the decision of what to benchmark can be a highly subjective one. Frequently a customer complaint, an employee suggestion, or simply a perception that a particular process is underperforming leads to a benchmarking exercise. Whether the exercise is successful and contributes significantly to business results, is very much left to chance. A more scientific method of process selection should be employed to increase the certainty of a successful benchmarking exercise.

As mentioned earlier, there are a number of factors that influence benchmarking priorities. For all organizations the issues regarding a process's or function's future strategic importance (i.e. potential business leverage) are particularly important. The other process parameters that are worth considering for benchmarking will vary from industry to industry and from one company to another.

2.6.8 The prioritization matrix: a tool for process selection in D2D

An organization such as D2D provides a good example of selecting appropriate process parameters. It is operating in a high-technology market, where innovation is particularly important (innovation is widely recognized as a process that is difficult to benchmark), and has limited experience of process-driven benchmarking. Therefore, the most applicable other parameter for a company such as D2D must be the ease with which a process or function can be benchmarked.

In establishing its total quality culture D2D has translated its strategic direction into a mission, critical success factors and critical processes, and has also developed an extensive performance measurement system. As the critical processes are the processes that have been identified as those most important to the company's strategic direction, these must be considered high priority for benchmarking.

To assess the listed critical processes against the two parameters identified as particularly relevant to D2D and its competitive situation, two questionnaires where developed for completion by D2D managers. One related to strategic importance issues, and the other to ease of benchmarking issues. A sample of the questionnaire design is shown in Table 2.2.

The questionnaires had a total of seven questions, each with equal weighting. The questions asked, with reference to the critical processes strategic importance and ease of benchmarking, are given below:

Strategic importance

Question (A) Relative to each other, how much does each process contribute to business results? (1 = Low/little contribution: 5 = high/larger contribution.)

Table 2.2 *Sample of the strategic importance questionnaire completed by D2D managers*

Question (A) *Relative to each other, how much does each process contribute to business results?* (1 = Low/little contribution: 5 = High/larger contribution) Rating (Please tick)	**1**	**2**	**3**	**4**	**5**
System, product & service delivery	–	–	–	–	–
Prospective customers	–	–	–	–	–
Competitive technology	–	–	–	–	–
People satisfaction	–	–	–	–	–
Process improvement	–	–	–	–	–
Supplier partnerships	–	–	–	–	–
Cost reduction	–	–	–	–	–
Self-assessment	–	–	–	–	–
Recognition	–	–	–	–	–
Deployment	–	–	–	–	–

Question (B) Relative to each other, how much does each process contribute to customer satisfaction? (1 = Low/little contribution: 5 = high/larger contribution.)

Question (C) What is the comparative future potential of each process? Relative to the other critical processes, how significant would the opportunities be if this process was dramatically improved? (1 = Low comparative potential: 5 = large comparative potential.)

Question (D) What is the relative contribution of each process to D2D's competitive advantage? (1 = Low/little contribution: 5 = high/larger contribution.)

Question (E) What is the current performance of each process, either in comparison to external benchmark data or from your own perception? (1 = High/good performance: 5 = low/poor performance.)

Question (F) By comparing the processes, what is the relative threat to D2D's competitive position if this process is not improved? (1 = Relatively low/insignificant threat: 5 = relatively high/significant threat.)

Question (G) Currently, what is the resource requirement of each process, i.e. capital, labour, time, etc.? (1 = Low/small resource requirement: 5 = high/large resource requirement.)

Ease of benchmarking

Question (A) What is the level of TQ implementation in the process? i.e. is the process fully mapped, are the process owners fully trained, are performance measures kept, do quality circles meet regularly, etc.? (1 = Low/little TQ implementation: 5 = high/significant level of TQ implementation.)

Question (B) Would you consider the process to be generic in nature (i.e. do many other organizations carry out a similar process) or very specific to D2D? (1 = Specific to D2D: 5 = highly generic.)

Question (C) Is improving the process relevant only to a few individuals, or many groups throughout D2D? (1 = Only a few individuals: 5 = organization-wide.)

Question (D) How easy do you consider it to be to identify benchmarking partners for this process? (This can include companies not in the same industry.) (1 = Difficult to identify partners: 5 = partners easily identifiable.)

Question (E) How readily available do you consider information on the process to be (From both D2D and any benchmarking partner)? (1 = Information difficult to obtain: 5 = information readily available.)

Question (F) How willing to share information on the process do you consider D2D to be? (1 = Not willing at all: 5 = very willing to share.)

Question (G) What is your perception of the opportunities for improving the process, i.e. is further improvement of the process likely to be easy or difficult in future? (1 = Easy to improve/many opportunities: 5 = difficult to improve/few opportunities.)

Ideally these questionnaires would be completed by consensus agreement among the senior management team or quality council. However, due to the time limitations of this project, the questionnaires were completed by managers as individuals. In practice, in a total quality organization with well-communicated strategies by an established goal-deployment process, it is likely that the answers managers give to the questions will not vary greatly.

It may even be the case that non-TQ organizations would find some differences of opinion between their managers, as the strategic direction and processes are not so well understood. There may, in fact, be a relationship between the difficulty an organization has in completing the questionnaires and how well understood the mission and processes are.

The strategic importance questionnaire itself is a tool by which a senior management team can consider the strategic implications of the critical processes and how much value is added by each. It provides an opportunity to reassert the importance of the various critical processes and gain commitment, not only to the Benchmarking exercise selected but also to the organization's strategic direction in its entirety.

Although it is the overall score that the critical process receives that is relevant to the prioritization (a ranking is not necessary), some managers found

Table 2.3 *Results of the Strategic Importance questionnaire*

Process	*Question A*	*Question B*	*Question C*	*Question D*	*Question E*	*Question F*	*Question G*	*Total*
System, product & service delivery	5	5	5	4	2	5	5	31
Prospective customers	3	3	4	4	3	4	4	25
Competitive technology	4	4	5	4	2	4	4	27
People satisfaction	3	3	3	2	4	4	3	22
Process improvement	4	3	3	3	3	4	3	23
Supplier partnerships	2	2	3	4	2	3	2	18
Cost reduction	3	2	5	4	3	4	3	24
Self-assessment	3	3	3	3	3	3	3	21
Recognition	1	2	3	3	2	2	2	15
Deployment	2	4	4	3	3	3	2	21

that paired comparison was a useful technique to help in the decision-making process. The scores for strategic importance and ease of benchmarking, derived from the questionnaires are given in Tables 2.3 and 2.4.

Having rated the Strategic Importance and Ease of Benchmarking processes, the results can be related to one another. Because the two measures are independent, one would expect some processes to score high on Strategic Importance and low on Ease of Benchmarking whereas others will score high on Ease of Benchmarking and low on Strategic Importance.

Because of this independence, it would be meaningless to add the scores together, or to look at the characteristics independently. There is however, considerable value in plotting the values of each on a matrix with Strategic Importance on one axis and Ease of Benchmarking on the other (see Figure 2.8).

As the matrix is a relative comparison of the critical processes, the axis can simply range from the lowest to the highest scores of each parameter. In the case of D2D's critical processes:

- Strategic Importance will range from **15** to **31**
- Ease of Benchmarking will range from **13** to **31**.

The best opportunities for a successful benchmarking exercise are obviously those processes that score high on both characteristics, and therefore occupy the top-right sector of the matrix. Benchmarking exercises carried out on

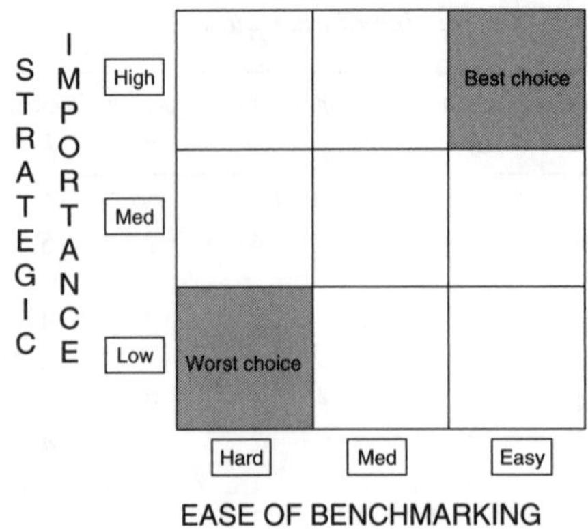

Figure 2.8 *Critical process benchmarking prioritization matrix*

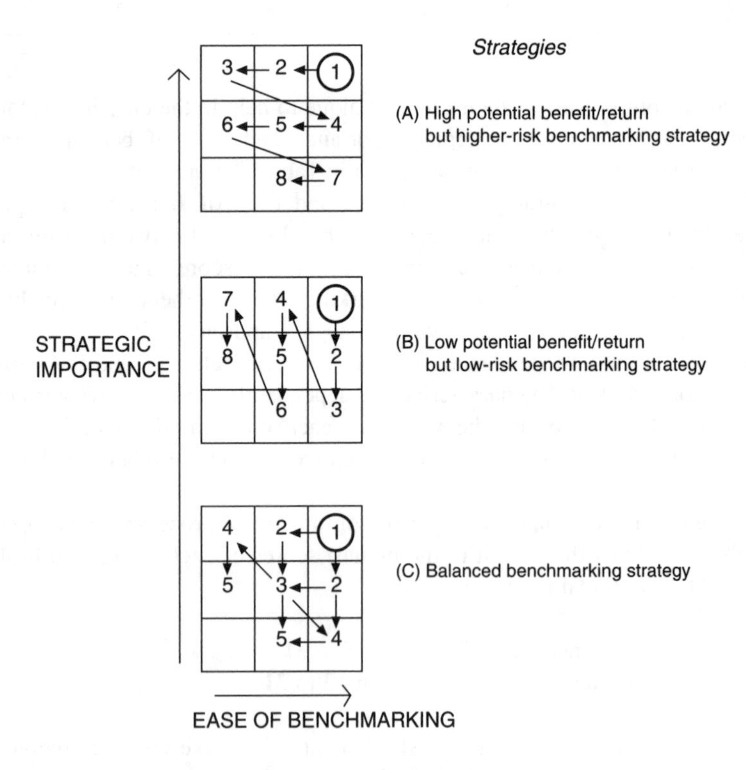

Figure 2.9 *Selection of critical processes based on alternative benchmarking strategies*

processes in this sector are more likely to yield reliable results fairly easily, and any performance improvement in the process is likely to have a significant beneficial impact on the overall performance of the organization.

The selection of processes from elsewhere within the matrix then depends on the benchmarking strategy the company selects (see Figure 2.9). If it wishes to pursue a high potential benefit strategy, i.e. a strategy that benchmarks processes of high strategic importance, but pays little attention to the ease with which that process can be benchmarked (a high-risk strategy), it would select the processes falling into the categories in the order of preference shown for strategy A. If the company wishes to minimize risk and pursue benchmarking projects that are more likely to produce successful (but not so impressive) results, it would select processes in the order shown by strategy B. If it wishes to select processes for benchmarking on the basis of both measures – Strategic Importance and Ease of Benchmarking – it would select processes in the order shown for strategy C.

Which strategy is selected will largely depend on the organization's expertise in benchmarking, the nature of its industry and the distribution of critical processes within the matrix. These issues are considered with reference to D2D's critical processes. Figure 2.10 shows the results of the benchmarking questionnaires (see Tables 2.3 and 2.4) plotted on the prioritization matrix.

From the matrix it can be seen that the critical process of system, product and service delivery occupies the top-right sector of the matrix and is, therefore, the best choice for process-driven benchmarking. One of the important characteristics of the matrix, as a process-selection tool, is that mapping the location of the processes on the matrix enables managers to visualize how they are positioned in relation to the two parameters. This should act to gain

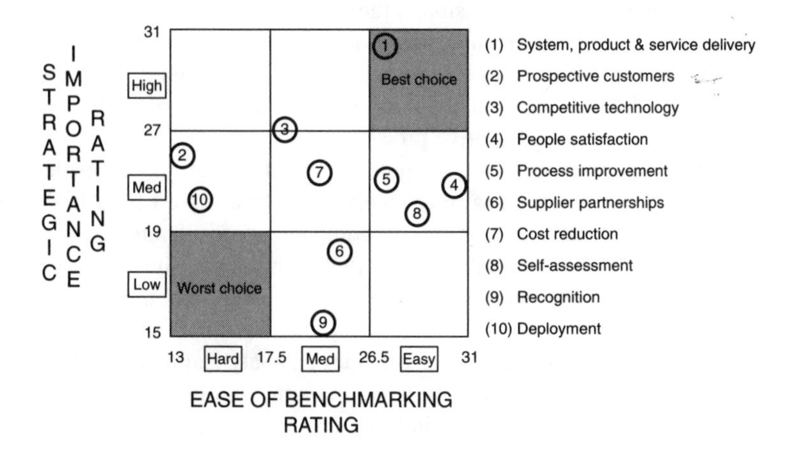

Figure 2.10 *The prioritization matrix applied to D2D's critical processes*

Table 2.4 *Result of the Ease of Benchmarking questionnaire*

Process	*Question A*	*Question B*	*Question C*	*Question D*	*Question E*	*Question F*	*Question G*	*Total*
System, product & service delivery	4	5	4	3	5	3	3	27
Prospective customers	3	4	1	1	2	1	1	13
Competitive technology	3	3	2	2	3	1	4	18
People satisfaction	4	5	5	5	5	4	3	31
Process improvement	3	5	4	4	5	3	3	27
Supplier partnerships	4	4	4	3	2	4	4	25
Cost reduction	3	4	3	1	5	2	4	22
Self-assessment	4	5	5	3	5	5	2	29
Recognition	4	2	2	2	5	5	3	23
Deployment	2	2	2	1	5	1	2	15

commitment to both the first-choice benchmarking project and the overall benchmarking strategy.

Due to the distribution of D2D's critical processes on the matrix (three critical processes medium in Strategic Importance and Easy to Benchmark), plus the fact that they are relatively inexperienced in process-driven benchmarking, a low-risk/balanced strategy would appear most appropriate. Another consideration that supports this conclusion is the nature of the competitive environment in high-technology manufacturing, i.e. the difficulty of benchmarking increases rapidly from right to left on the matrix. This is mainly the result of the problem in identifying benchmarking partners and lack of willingness to share information about innovative processes.

An organization in a lower-technology industry that has considerable experience of benchmarking is most likely to be best served by a high-risk benchmarking strategy. However, for all companies, consideration of the distribution of their processes on the matrix, their competitive situation and their experience of benchmarking is necessary for selection of the appropriate benchmarking strategy.

The prioritization matrix therefore provides an analytical method of assessing the critical processes against the benchmarking priorities that are considered pertinent to the organisation's industry and experience of benchmarking. Figure 2.11, summarizes the construction of the various factors and scores that make up the matrix.

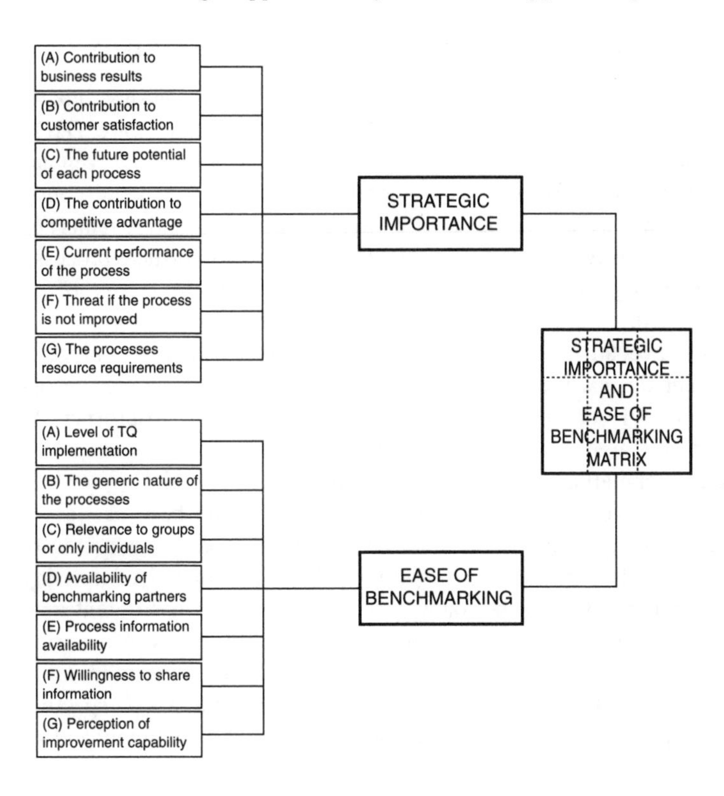

Figure 2.11 *Analytical system for the selection of critical processes for benchmarking*

In a total quality organization such as D2D the matrix should not be difficult to complete as the strategic direction and the critical processes themselves are well understood. The matrix provides a systematic procedure for selecting the critical process to be benchmarked. It helps to remove individual bias by rationally considering both the strategic direction (without reference to frequently misunderstood terms such as 'value chain', 'value-added' or 'core competencies') and the difficulty of benchmarking that process.

A critical factor for successful benchmarking is the commitment of senior management. As the matrix is a highly visible representation of a selection procedure that they have participated in, they are more likely to commit themselves to both the initial benchmarking project and the benchmarking strategy. For a total quality company, such as D2D, that has yet to undertake a process-driven benchmarking exercise independently of consultants, the matrix provides a useful framework by which to prioritize their future benchmarking projects.

2.6.9 Process-driven benchmarking: the next step for D2D

For D2D process-driven benchmarking represents a missing link in their journey towards the standards of excellence to which they aspire. D2D's main benchmarking emphasis to the present has been on a results-driven approach using the databases and recommendations of management consultants. Although this is a useful introduction to benchmarking, and remains appropriate for competitive benchmarking, D2D are in a position where process-driven benchmarking could yield significantly benefits in terms of process improvement.

D2D have an established total quality culture where the processes are documented, measured and continuously improved. These are the optimum conditions for continuous process-driven benchmarking to prosper. As D2D are part of a large multinational organization the opportunity to begin process-driven benchmarking internally should also be considered.

Process-driven benchmarking extends the search for best practices beyond immediate competitors to consider organizations from various industries, to reach for the best of the best performance levels. Frequently, surprising similarities in processes can be found between companies in different industries. By adapting appropriate ideas from best practice, regardless of its origin, into the organization's own processes a significant performance improvement can result. Process-driven benchmarking enhances the external perspective of an organization's total quality programme and is, therefore, a timely and valuable quality improvement tool for D2D to employ.

D2D's total quality culture represents the ideal environment for process-driven benchmarking to significantly improve the processes that are critical to their success. By introducing such benchmarking, D2D will ensure that they know the performance levels needed to be world competitive and understand the processes by which this can be achieved.

2.7 Strategic benchmarking: key benefits

Strategic benchmarking provides the opportunity to take a 'helicopter view' of the business and take action where and when necessary. It is the best means for building synergy levels and integrating the various key elements of the business. The model illustrated in Figure 2.12 represents an integrated approach to modern competitiveness facilitated by the practice of strategic benchmarking:

- The model itself defines a new approach to marketing where the traditional **4Ps** of Price, Promotion, Place, Product are replaced with a set of new Ps standing for:

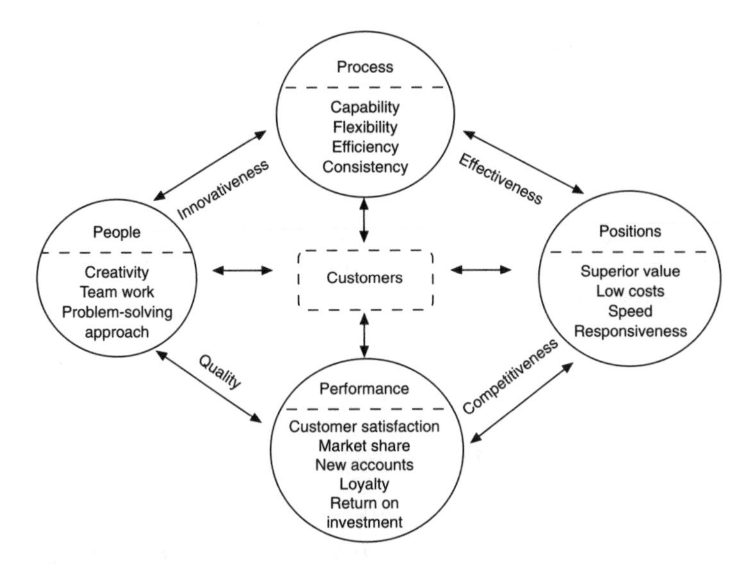

Figure 2.12 *Modern marketing – an integral part of competitive strategy*

PROCESS
PEOPLE
PERFORMANCE
POSITION

- At the heart of strategic planning is the *customer*. Strategic benchmarking ensures that corporate priorities have to be defined in customer terms.
- Corporate goals and targets have to be measured in terms of:
  – Quality (defined from the customer perspective)
  – Effectiveness (value-added contribution)
  – Innovativeness (uniqueness, creativity and competitive advantage)
  – Competitiveness (market position and domination)
- The 4 Ps are the critical building blocks and represent the following:
  – Process represents optimized capability and ability to respond to changes in the marketplace
  – People are the major asset and the real source of value-added contribution through creativity and innovation
  – Performance needs to be measured in terms of customer satisfaction, building long-term partnerships and new acquisitions
  – Position is the arena of 'competitive performance', and needs to be measured in terms of superiority from different parameters

Strategic benchmarking ensures that:

- **Focus** is always on the customer and market
- **Corporate goals** are clearly defined and based on internal/external benchmarking analysis
- **Practices** of people and processes are state of the art and pioneering
- **Performance gaps** are measured from customer/market perspective
- **Results** are monitored to ensure that there is a positive impact from the *integrated* approach to modern management.

References

1 Watson, G.H. (1993) *Strategic Benchmarking – How to rate your company's performance against the world's best*, John Wiley, New York.
2 Hutton, R. and Zairi, M. (1994) D2D: A quality award winner's approach to Benchmarking, *International Journal of Benchmarking Quality Management and Technology* **1**, No. 3, pp. 21–38.

3 Partner selection

Less is more

Robert Browning

The best is good enough

German proverb

Whatever suffices is enough

Latin proverb

3.1 Partners or competitors?

The question of choosing the right partners to benchmark against is invariably one of the most difficult areas to handle. If the notion of 'industrial tourism' is to be avoided then it follows that benchmarking partners have to be selected for the right reasons, the visits to be conducted have to be legitimately assessed and agreed upon. Ease of access is not necessarily a good reason for agreeing to visit a certain organization. The selection of benchmarking partners as addressed in this chapter is related to the specific areas of interest (i.e. *what to benchmark*). Different world-class organizations could be of interest because of certain *core competencies* they may possess. For instance:

- The effective implementation of total quality management: Rank Xerox, IBM, Hewlett-Packard, etc.
- Supply-chain management: fast-moving consumer goods sector, retail sector, electronics industry, etc.
- Customer service and customer satisfaction: airline Industry, automobile industry, computer industry, etc.

The selection of a benchmarking partner has to be arrived at from the investigation of the process under investigation and not at the onset of the benchmarking project. In many instances, when the process itself is broken down into critical elements it may be the case that learning will have to come from a blend of partners and that the issues are quite diverse to be addressed through the one single visit. The consequence of poor partner selection is poor results, misguidance and detrimental effects on the improvement strategy to be recommended.

3.2 The breadth and depth of benchmarking

Benchmarking offers a wide variety of opportunities and can be used flexibly for many different perspectives. First of all, *what type of benchmarking we use* will determine ultimately what kind of information will be sought and where the source of obtaining it is going to be determined:

- **Strategic benchmarking**, for instance, is a process which is used for identifying world-class standards, determining gaps in competitiveness, developing relevant strategies and generally speaking for remaining focused and being aware of developments in the marketplace. This may require not visits to partners but some high-level studies.
- **Process benchmarking** is where the bulk of benchmarking activity takes place, it is about finding out *how* good performance is achieved and learning about best practices in order to put together action plans capable of closing any likely competitive gaps. Very often, process benchmarking necessitates visits to partners and thorough preparation is required.
- **Performance benchmarking** is an important process for establishing 'benchmarks' and identifying what 'stretch objectives' need to be put in place. Very often, this approach may not require company visits and is conducted through consortia studies and with the facilitation of consultants and third-party involvement.

It has to be said that particularly for strategic benchmarking and performance benchmarking, the involvement of key competitors is very often to be expected. Further, and in the case of all three types of benchmarking, a 'first stop' may have to be *internal benchmarking* with sister divisions, particularly in the case of large corporations.

When one talks about *competitive benchmarking* there is a thin dividing line between legitimate, public domain-based information and industrial espionage. Archer [1] indicates that there is a wide spectrum of information which can be obtained either easily and legitimately or through devious means or only legitimately through the practice of benchmarking (Figure 3.1). A lot of information can be gathered from company reports, market research, surveys, etc. Very often, information relating to innovation and pricing strategy is much harder to gather.

In fact the practice of benchmarking is somewhat determined by the quality information sought and therefore the level of preparedness to incur costs. As Figure 3.2 indicates, general industrial surveys are relatively cheap but

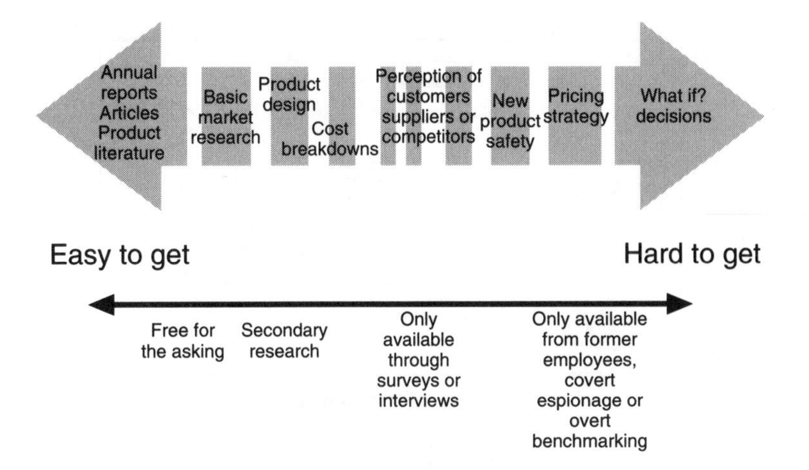

Figure 3.1 *The spectrum and ease of obtaining information*

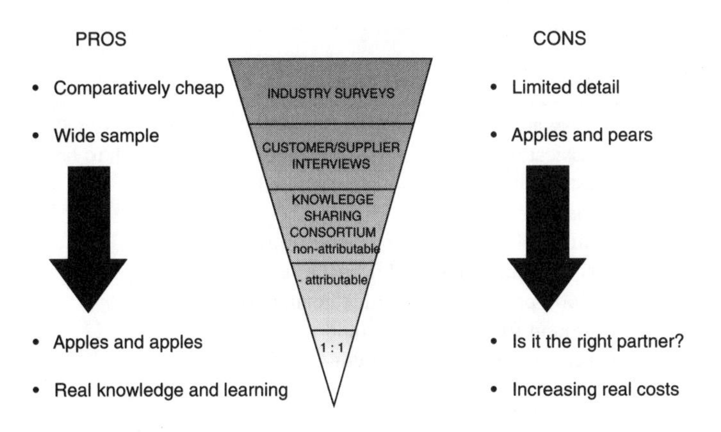

Figure 3.2 *A comparative analysis of information quality*

they offer crude and limited information. When the process becomes much more focused and very attributable to provide an 'apple versus apple' scenario, the value goes higher and therefore so does the cost [1].

There is therefore a general rule in that the value of information is directly related to the difficulty in accessing 'good organizations' as Figure 3.3 illustrates. While internal benchmarking can have a wide variety of benefits, it does suffer from limitation. Generic benchmarking, on the other hand, offers the best venue but has different issues associated with it. Table 3.1 contrasts the various approaches.

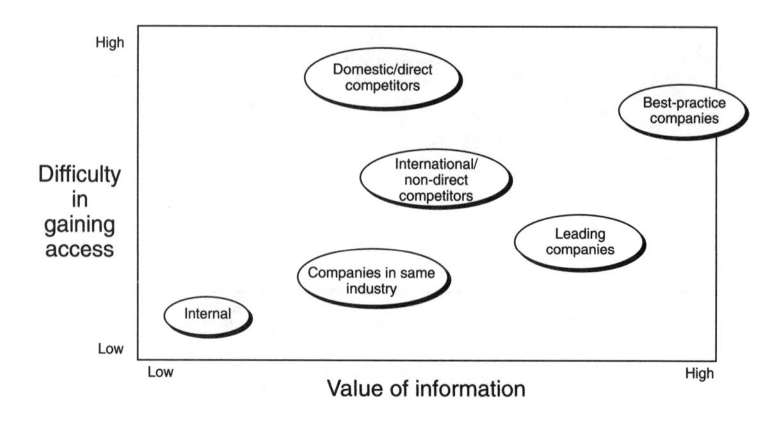

Figure 3.3 *Access versus value of information*

Table 3.1 *Pros and cons of benchmarking choices*

Type of benchmarking	*Pros*	*Cons*	*Remarks*
Internal	• Easy access • Easiest way to establish a baseline • Identifies differences between sister companies • Good preparatory ground for external benchmarking activity	• Opportunities for improvement limited to internal best practice only • May be seen as threatening and could create sub-optimization • Only useful in the context of larger organizations	* Still a very valuable approach to be used
Competitive	• Helps prioritize areas of improvement according to the competition • Helps determine initial area of interest to most organizations	• Often limited pool of participants • Opportunities for improvement limited to 'known' competitive practices	* Difficult and limits ambitions

Table 3.1 *continued*

Type of benchmarking	*Pros*	*Cons*	*Remarks*
cont.		• Hard to recruit participants for detailed and reliable data sharing	
Industry	• Provides industry trend information • Provides management with a conventional basis for quantitative comparisons	• Quantitative data are difficult to obtain • Learning points limited to industry norms	* May encourage complacency
Best in class	• Examines multiple industries • Best option for the identification of radically innovative practices • Broadest perspective • 'Ultimate' goal orientation	• Often difficult to identify 'best in class' organizations • Findings may not be comparable 'Best in class' organizations besieged with requests . . . survey participation may be difficult to obtain	* Source of innovation and learning

3.3 The psychology of effective benchmarking

In many instances many benchmarking opportunities are hampered by human prejudices and personal biases. Very often the following factors are cited as the major obstacles encountered.

- Senior management lack of buy-in attributing their final decisions to halt benchmarking proposals to cost factors, as a low priority, or that benchmarking is not what is really needed
- The practice of 'we are unique', 'we are different' and therefore there is very little mileage for us to learn from others
- The unsystemized approach in selecting areas for benchmarking
- The poor selection of partners and perhaps sometimes the large number of

partners selected could unnecessarily escalate costs without enhancing the benefits by a great deal
- The time constraint put on the project means that the focus is diverted from ‘real learning opportunities’ and more towards ‘project completion’
- Lack of readiness to reciprocate and share information with partners.

There are, however, very good ground rules to subscribe to when doing bench-marking. In addition to the international code of conduct which is described separately, it is important to ensure that the following issues are addressed:

1 Partners need to be told at the onset that their selection was because of hard factual information and not ‘hearsay’. In other words, the learning potential has been assessed and identified.
2 There is definitely a need to demonstrate to partners that they stand to benefit from the exercise and that the feedback will be of great value to them.
3 The purpose should also be to put in place the foundation for a long-term partnership by perhaps demonstrating synergies, similarities and sensitivities to each others’ main concern.
4 The process of benchmarking needs to be clearly demonstrated to partners (i.e. what, why, where, who, what outputs, what timescale, what resource requirements etc.).
5 Agree on formats and performance measurement strategy so that outcomes are comparable and benchmarks can be established.
6 Work out the reciprocation strategy by clearly identifying what the partner’s interests are. It could well be that they would like to tackle a different area from the one suggested to them.

3.4 Best practice approaches in partner selection

At the heart of partner selection is really the importance of understanding the processes which need to be benchmarked. This is reported time and time again as a key prerequisite and is often the first step in launching a benchmarking project. A typical approach widely recommended is performing a strategic review of key processes and gauging their importance in delivering critical success factors and their degree of effectiveness. This is covered in another part of the book. But nonetheless the broad generic methodology for this is to relay the importance of each project against its effectiveness and all those critical processes which fall short of being truly competitive can be considered as contenders for benchmarking (Figures 3.4 and 3.5).

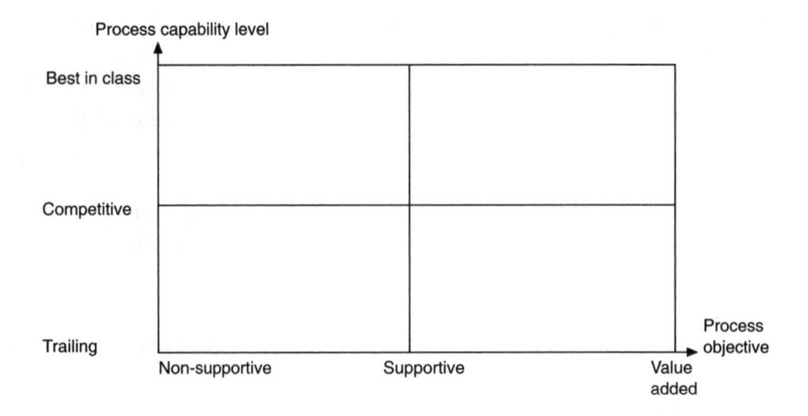

Figure 3.4 *Process selection matrix*

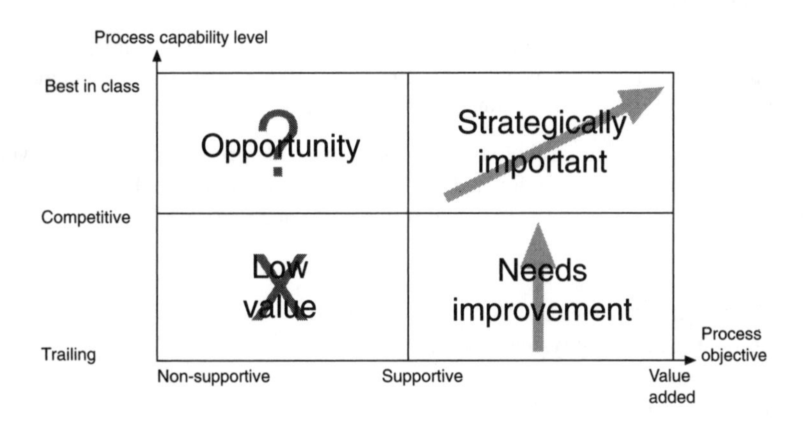

Figure 3.5 *Selecting opportunities for benchmarking*

3.4.1 Rover Cars approach

Within Rover Cars, partners are selected on the following basis:

- **Internal operations**: similar functions within different operating units performing the same tasks. Rover recognizes that this step is necessary, even if the processes are not of world-class standard. Sharing of internal information is thought to be a useful step before going outside.
- **Direct product competitors**: This is encouraged as an option to consider, even though it is recognized that access may be difficult and complications are necessarily likely.

- **External non-competitors**: For generic processes such as logistics, accounting, recruitment, etc. Rover teams are encouraged to consider this route, which is the least threatening and where 'genuine innovative solutions' can be found. Appendix 3.1 illustrates a proforma used for classifying likely benchmarking partners.

Within Rover Cars, benchmarking teams are encouraged first to carry out initial research prior to directly contacting potential partners. This is thought to be important for two reasons:

- To understand the performance of the process in question in terms of *metrics*
- To understand exactly how the process works (*practices*).

A proforma for this purpose is illustrated in Appendix 3.2. This stage is so critical since it provides a focus on the selected partners. As stated in the Rover benchmarking guidelines:

> It should be emphasized that a great deal of the required information can be obtained at the 'desk research' stage. Direct contact should be used to fill in the picture, not to get initial information about companies.

In addition to a comprehensive methodology and specific guidelines Rover has the following prompts for its benchmarking teams to make the experience very worthwhile. Comparisons need to take into account:

- Methods involved
- Policies and procedures affecting the method
- Geographical and physical layout
- Management responsibilities
- Skills and training levels
- Investment and operating costs
- The relationship between the specific benchmarked process and the rest of the business – how it fits into the big picture.

3.4.2 IBM approach

Like the Rover approach, IBM encourages teams to use a wide variety of information sources before considering potential partners. As far as potential partners are concerned, the following sources need to be considered:

- Companies that have won quality/business awards
- Top-rated firms in industry surveys
- Success stories published in periodicals

- Statements of pride in business articles, such as 'we have a world-class reservation system', 'our business has continued to grow well ahead of competition', etc.
- Companies with excellent financial results
- Feedback received from internal and external experts, customers, suppliers and business partners.

In order to arrive at a comprehensive and effective list of benchmarking partners, IBM uses a methodology based on a matrix which matches potential benchmarking partners to best-of-breed or world-class requirements. Through a questionnaire the results are plotted on the matrix and the best performers according to the set criteria can then be considered further as potential partners. Some of the key questions which can be used in this process, include the following:

- For what quality process or results is the company especially known?
- What is your evidence that it is an industry leader in the area of interest?
- What is the company's revenue?
- What is the company's profit?
- What is the company's market share?
- What is the company's level of customer satisfaction?
- Has the company made any contributions to the state of the art in its industry through papers presentations?

The internal process for preparing for benchmarking includes the following key steps:

1 Compiling the characteristics (distinctive elements) of the process or business area.
2 Listing best-of-breed or world-class criteria for the business area or process of interest.
3 Listing of currently inhibiting factors for the process or business area to fulfil their function.
4 Preparing the open-ended questions for assessing best-of-breed and world-class factors.
5 Developing a list of benchmarking partners.

3.4.3 American Express approach

The American Express approach to benchmarking is very thorough and comprehensive. It has some key characteristics which are worth considering:

- A system for monitoring in- and out-bound benchmarking requests

- A system for preparing the teams prior to official company visits taking place.

It is recommended for all benchmarking project teams to prepare some background information on areas of interest, with a proper process map, an indication of customers/suppliers and performance. Data may cover cycle time, cost, quality, etc. Further, an indication is required of who is involved in the benchmarking expedition, what questions are likely to be asked and finally, a confidentiality agreement (Appendix 3.3).

In order to close the loop and provide useful feedback to partners' visits, a system has been devised to encourage teams to immediately provide feedback on the outcomes of the visit and any likely prompts, advice, help and support which may be useful to the partners concerned (Appendix 3.4).

3.4.4 Rank Xerox approach

A very useful roadmap is used within Rank Xerox (Figure 3.6). In order to select the most appropriate benchmarking partners:

1 The key trigger for selecting partners is the process in question, and in the case of competitors, product range is a very applicable criterion.
2 On the question of a competitor, if the match is perfect and the exchange is going to be strategically beneficial, the final selection is going to be determined by anti-trust laws and legal and ethical factors.
3 If it is a non-competitor organization, the factors which may rule out the selection are differences in culture and in the criteria for benchmarking and, generally, lack of 'synergy' and 'chemistry'.
4 As above, some partners could still be a low-priority choice because even though a cultural affinity may exist, organizational differences may automatically rule them out.
5 The best category for selection are those partners that have not only analogous processes and similar organizational types, but also similar cultural affinities. Further, they fit the criteria that Rank Xerox desires and are committed to a full exchange.

The issue of partner selection therefore is not straightforward. While ease of access is very tempting and attractive, nonetheless it could be a reason for justifying 'industrial tourism' and wasting of resources. On the other hand, difficulty of access should not deter teams from pursuing what is important for the business, what is right to do and if the project team subscribes to all the principles covered in this chapter, most partners targeted would feel privileged and happy and therefore obliged to share information on a win–win basis.

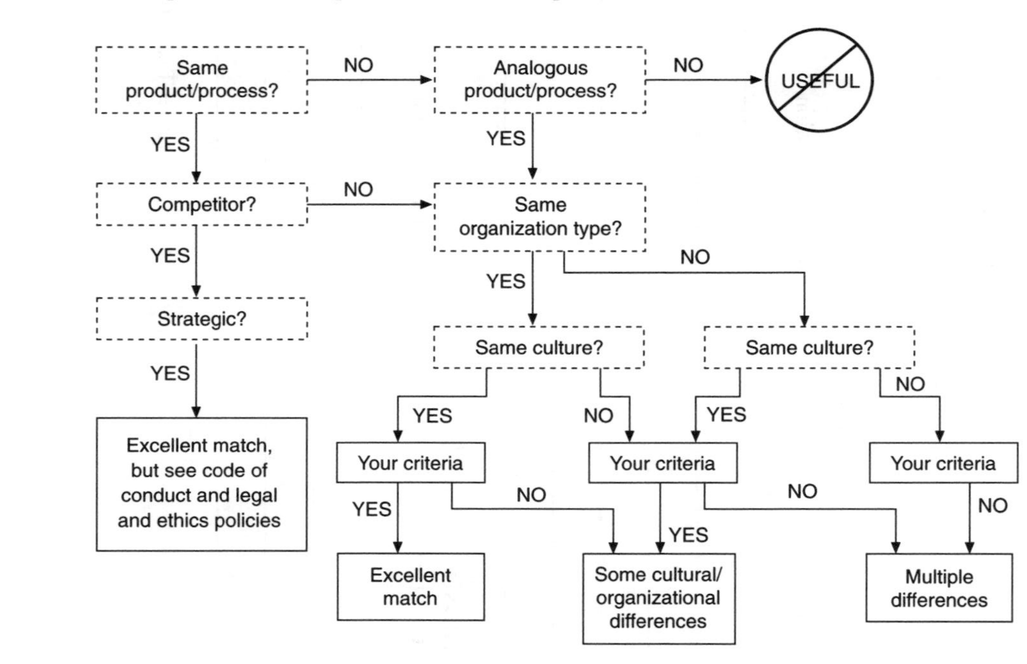

Figure 3.6 *Rank Xerox partner selection roadmap*

Appendix 3.1: Rover Cars approach to partner selection

ROVER INTERNAL:

(✓)	Potential partner	Function, location
☐		
☐		
☐		
☐		
☐		
☐		

DIRECT PRODUCT COMPETITORS:

	Potential partner	Function, location
☐		
☐		
☐		
☐		
☐		
☐		

EXTERNAL NON-COMPETITORS:

	Potential partner	Function, location
☐		
☐		
☐		
☐		

(✓) = Taken further

Appendix 3.2: Rover Cars approach to information collection

> Instruction: Develop a list of resources as per example. Assign responsibility for each and set time goals for the completion of all research. Tick each item as it is completed.

INTERNAL RESOURCE	WHO IS RESPONSIBLE	BY WHEN	(✓)
			☐
			☐
			☐
			☐
			☐
			☐
EXTERNAL RESOURCE			
			☐
			☐
			☐
			☐
			☐
			☐
			☐
			☐
			☐
			☐
			☐
			☐
			☐

Appendix 3.3: Checklist of benchmarking requirements at American Express

Recommended benchmarking requirements checklist

Prior to conducting a benchmarking study with an outside organization, the following information should be available from American Express. Check all that are available.

- High-level, cross-functional description or map of the area to be benchmarked, including:
  Customers
  Suppliers/vendors
  Inputs/outputs
  Departments/areas impacted.
- Detailed flowchart(s) of specific area(s) to be benchmarked.
- American Express performance data, including, but not limited to:
  Cycle time measures
  Cost measures
  Quality (e.g., error rates, customer complaint information, etc.)
  Customer measures.
- Agenda for visit
  Benchmarking participants from American Express
  List of questions to be covered in visit
  Non-Disclosure Agreement.

Appendix 3.4: Benchmarking site visit report proforma at American Express

Outbound benchmarking or site visit information
(To be filed after American Express benchmarks or conducts a
site visit at another company)

Date ____

American Express contact name ______________________

Title ____

Department ____

Address ____

____

Phone ___________________ Fax ______________________

Company benchmarked or visited ______________________

Contact name _______________ Phone __________________

____

City ____________________ State ____________________

Primary purpose:

Benchmarking ______________ Site Visit _______________

Brief description of general area or process reviewed:

____

____

____

____

____

Dates data collected ____________________________________

Abstract of results (1–2 paragraphs) or attach trip report. For benchmarking projects, please indicate information available to other Amex Departments/Business Units.

____

____

____

____

____

____

Reference

1 Archer, R. (1996) Tackling the secrecy and sensitivity issue: The ethics of benchmarking and its International Code of Conduct, 1st National Symposium on Best Practice Management, Kuala Lumpur, November 1996, Malaysia.

4 The ethics of benchmarking: how to handle secrecy and sensitive issues

I have studied the enemy all my life. I have read the memoirs of his generals and his leaders. I have even read his philosophers and listened to his music. I have studied in detail the account of every damned one of his battles. I know exactly how he will react under any given set of circumstances, and he hasn't the slightest idea of what I'm going to do. So when the time comes, I'm going to whip the hell out of him

George S. Patton
American General

Not everything that counts can be counted; not everything that can be counted counts

Albert Einstein

Nothing is good or bad but by comparison

Thomas Fuller

To compare is not to prove

French proverb

4.1 The overt nature of benchmarking

Benchmarking is not a means of winning at any cost. It is a legitimate, systematic, overt and ethical process of bringing about effective competitiveness. It is driven by the spirit of 'learning through sharing' and certainly does not represent an obsession with competitors. The myth that you can only compete well if you know what your competitors are doing and you have to be able to copy what they do to succeed, has been broken. Learning is generic, management is universal and best practice is not a privilege that only a few organizations have.

Unlike industrial espionage, benchmarking is in an overt fashion, a win–win principle. It is concerned more with finding out about ideas on managing processes and therefore achieving superior performance rather than with gathering sensitive information on cost, pricing, etc. which, very often, tends to be the subject of anti-trust law. After all, anti-trust and anti-competitive legislation was brought about in the first place to:

- Encourage and promote high quality
- To ensure there is a variety
- Low prices
- Innovation and efficiency.

There is therefore a clear dividing line between benchmarking and industrial espionage (Table 4.1). Industrial espionage is a deliberate, unethical, devious, obsessive way of obtaining information at any cost. It is about breaking the rules of fair competition, wanting to sabotage, perhaps drive competitors out of business and even to mislead customers in order to create a win–lose scenario. The 'backdoor' principle of finding out about competitors is not synonymous with benchmarking. Those advocates of unethical principles who claim that benchmarking is no more than a cleverly packaged approach to industrial espionage are totally mistaken and do not really understand the way benchmarking works.

Furthermore, as has been discussed time and time again, benchmarking does not represent an obsession with competitors. There are more than enough reasons to justify that an obsession with competitors is not perhaps the best way of enhancing competitive performance:

1 **The late factor:** information on competitors coming from them could be 'poisonous' and very misleading. It could be a deliberate attempt to foil plans, sidetrack competitors or totally mislead them.
2 **Unreliability of sources:** The quality of the information itself may be in question. Since it is obtained through 'devious means', second or third hand, it could be open to questioning or validity and reliability.
3 **Different may not necessarily mean better:** the most dangerous signal is the fact that competitors could be found to be doing things 'differently'. This therefore could lead to an organization abandoning its own plans and

Table 4.1

Benchmarking	*Industrial espionage*
• Overt	• Covert
• Ethical	• Unethical
• Reliable	• Unreliable
• Identifies the key drivers of performance	• Is subject to hypothesis and guesswork

strategy and embarking on initiatives to copy what competitors are doing. There is always a perception that if competitors are doing something differently, this therefore means it is better. Of course, without any specific evidence (which is very hard to obtain) it should not be accepted that change for the sake of change is the best way forward.

4 **Industry best not good enough:** in many cases, industry best standards are not really world class. Competitors could become 'lethargic', hit by inertia and desperate for big innovations. If therefore benchmarking is to greatly assist by delivering the desired breakthroughs, some lateral thinking needs to take place by breaking the mould and looking outside the industry. The literature is flooded with examples of benchmarking learning and Eureka phenomena, which have assisted many organizations in making the gigantic leaps forward. The best quoted example is perhaps the classical benchmarking experience of Rank Xerox with L. L. Bean.

Competitive benchmarking is, however, still very possible as an ethical, overt, reliable way of learning from and about competitors. It takes place in a variety of ways:

Public domain reports, papers, books, etc.
Involvement in industrial associations
Participation at conferences, symposiums, etc.
Third party facilitated consortia
Role of consultants
Participation in major studies
Business schools.

4.2 Competition law

It is vital for any organization wanting to thrive and compete globally to understand what governs modern competition, what the rules of the game are and to ensure that there is a total adherence to the various laws in place, both at national and international levels. Whether as providers of goods and services, it is incumbent on every business to subscribe to the 'clean etiquette' principle. Failing to do that can mean swift reaction from the courts, severe penalties, reprimands for anti-competitive behaviour, plus the tarnishing of corporate image through poor media coverage.

As defined in one of the guidance manuals on competition law, produced by the Post Office in the UK, competition law is a set of rules that regulates business contacts to avoid unfair agreements or oppressive conduct. These unfair agreements and oppressive conduct can fall into two main categories:

1 Agreements between competitors that damage customers, other competitors, distributors or suppliers – for example, by raising prices, and
2 Monopoly conduct that allows companies to exploit their market power – for example, by trying to put small competitors out of business.

Contrary to what people might think, competition law is not restrictive to good business practice. It rather guides actions and practices in the right direction and avoids embarrassment and problems.

- Competition law can help warn senior managers of any possible infringements of the law and give possible guidance on how to deal with uncertainties.
- Through a continuous process of monitoring compliance, problems can be picked up very early and dealt with very efficiently.
- As far as authorities handling competition law issues are concerned, a business that adheres to the principles and guidelines set by legislation will be treated leniently in the event of a problem occurring.
- Competition law sets the ground for long-term, effective, deserving, ethical and sustainable competitiveness through model behaviour. Business taking high-risk short cuts, acting unprofessionally and driven by short-term gains will, in the long term, be caught and severely punished.
- Competition law can be used proactively to break into new markets or product areas or even open up anti-competitive cartels. Regardless of organizational size, the degree of abundance in resources, the type of market, industrial sector, etc. competition law protects the interests of all and is a constant reminder to those business organizations wanting to bully small competitors, play foul, try to cheat and abuse the rules of the game.

4.3 Sanction and implications for contravening competition law

Depending on the type of infringement, where it takes place, the nature and severity of the sanctions proposed can vary considerably.

4.3.1 UK sanctions

- Substantial fines
- Imprisonment of directors and employees for contempt of court for breaching a court order
- UK authorities can decide to control the future conduct of a business if its behaviour is thought to represent a breach of competition law. The wide range of powers that the authorities might exercise could, for instance, include:

  - An imposition of price controls
  - Ordering the sale of all or part of a business
- Agreements or arrangements which represent an infringement of the competition law could be invalidated and nullified.

4.3.2 European Commission sanctions

- Fines of up to 10 per cent of annual turnover can be imposed by the EC Commission if there are breaches in the law even if these are not intentionally carried out. This penalty could, of course, be very substantial in the case of large businesses.
- Similarly to the UK scenario, agreements can be invalidated if there is a breach of the law.
- Commercial contracts can be rendered illegal if there is non-adherence to competition law.

4.3.3 Other penalties

There are wider implications to be considered if the rules of the game are not played properly. Some of these could represent permanent damage to a business organization's reputation and could even mean its gradual demise. Among the various penalties to be considered are:

- The need perhaps to have to pay compensation to plaintiff companies thought to be harmed by illegal, unethical conduct
- The cost of legal suits, periods of uncertainty, commercial problems and psychological burdens
- Expense to compensate image through adverse press
- Having to be supervised, controlled and told what to do by the competition authorities.

Benchmarking and competition law go hand in hand. They are not contradictory and it is certainly not true to suggest that competition law can stifle the practice of benchmarking. It is perhaps more accurate to suggest that competition law and knowledge of the 'rules of the game' can assist in ensuring effective, ethical and fair outcomes from benchmarking projects.

In addition to understanding the process of benchmarking and the technical aspects of information exchanges, there are other principles that individuals and project teams involved in external activities need to know about. The International Benchmarking Code of Conduct was developed initially in the USA by APQC/SPI and a version of which was later developed for the European context is based on eight broad principles, covering:

1 The Principle of Preparation
2 The Principle of Contact
3 The Principle of Exchange
4 The Principle of Confidentiality
5 The principle of Use
6 The Principle of Legality
7 The Principle of Competition
8 The Principle of Understanding and Agreement

4.4 The European Benchmarking Code of Conduct

1.0 Principle of Preparation

1.1 Demonstrate commitment to the efficiency and effectiveness of benchmarking by being prepared prior to making an initial benchmarking contact.

1.2 Make the most of your benchmarking partner's time by being fully prepared for each exchange.

1.3 Help your benchmarking partners prepare by providing them with a questionnaire and agenda prior to benchmarking visits.

1.4 Before any benchmarking contact, especially the sending of questionnaires, take legal advice.

2.0 Principle of Contact

2.1 Respect the corporate culture of partner organizations and work within mutually agreed procedures.

2.2 Use benchmarking contacts designated by the partner organization if that is its preferred procedure.

2.3 Agree with the designated benchmarking contact how communication or responsibility is to be delegated in the course of the benchmarking exercise. Check mutual understanding.

2.4 Obtain an individual's permission before providing their name in response to a contact request.

2.5 Avoid communicating a contact's name in open forum without the contact's prior permission.

3.0 Principle of Exchange

3.1 Be willing to provide the same type and level of information that you request from your benchmarking partner, provided that the principle of legality is observed.

3.2 Communicate fully and early in the relationship to clarify expectations, avoid misunderstanding, and establish mutual interest in the benchmarking exchange.

3.3 Be honest and complete.

4.0 Principle of Confidentiality

4.1 Treat benchmarking findings as confidential to the individuals and organizations involved. Such information must be communicated to third parties without the prior consent of the benchmarking partner who shared the information. When seeking prior consent, make sure that you specify clearly what information is to be shared, and with whom.

4.2 An organization's participation in a study is confidential and should not be communicated externally without their prior permission.

5.0 Principle of Use

5.1 Use information obtained through benchmarking only for purposes stated and agreed with the benchmarking partner.

5.2 The use or communication of a benchmarking partner's name with the data obtained or the practices observed requires the prior permission of that partner.

5.3 Contact lists or other contact information provided by benchmarking networks in any form may not be used for purposes other than benchmarking.

6.0 Principle of Legality

6.1 If there is any potential question on the legality of an activity you should take legal advice.

6.2 Avoid discussions or actions that could lead to or imply an interest in restraint of trade, market and/or customer allocation schemes, price fixing, bid rigging, bribery, or any other anti-competitive practices. Don't discuss your pricing policy with competitors.

6.3 Refrain from the acquisition of information by any means that could be interpreted as improper including the breach, or inducement of a breach, or any duty to maintain confidentiality.

6.4 Do not disclose or use any confidential information that may have been obtained through improper means, or that was disclosed by another in violation of a duty of confidentiality.

6.5 Do not, as a consultant, client or otherwise pass on benchmarking findings to another organization without first getting the permission of your benchmarking partner and without first ensuring that the data are appropriately 'blind' and anonymous so that the participants' identities are protected.

7.0 Principle of Completion

7.1 Follow through each commitment made to your benchmarking partner in a timely manner.

7.2 Endeavour to complete each benchmarking study to the satisfaction of all benchmarking partners as mutually agreed.

8.0 Principle of Understanding and Agreement

8.1 Understand how your benchmarking partner would like to be treated, and treat them in that way.

8.2 Agree how your partner expects you to use the information provided, and do not use it in any way that would break that agreement.

Benchmarking protocol

- Know and abide by the European Benchmarking Code of Conduct.
- Have a basic knowledge of benchmarking and follow a benchmarking process.
- Companies should have:
  - Determined what to benchmark
  - Identified key performance variables to study
  - Recognized superior performing organizations
  - Completed a rigorous internal analysis of the process to be benchmarked before initiating contact with potential benchmarking partners.
- Prepare a questionnaire and interview guide, and share these in advance if requested.
- Possess the authority to share and be willing to share information with benchmarking partners.
- Work through a specified contact and mutually agreed arrangements.
  - When the benchmarking process proceeds to a face-to-face site visit, the following behaviours are encouraged:
- Provide a meeting agenda in advance.
- Be professional, honest, courteous and prompt.
- Introduce all attendees and explain why they are present.
- Adhere to the agenda.
- Use language that is universal, not one's own jargon.
- Be sure that neither party is sharing proprietary or confidential information unless prior approval has been obtained by both parties from the proper authority.
- Share information about your own process, and, if asked, consider sharing study results.
- Offer to facilitate a future reciprocal visit.
- Conclude meetings and visits on schedule.
- Thank your benchmarking partner for sharing their process.

Benchmarking with competitors

The following guidelines apply to both partners in a benchmarking encounter with competitors or potential competitors:

- In benchmarking with competitors, ensure compliance with competition law.
- Always take legal advice before benchmarking with competitors. (Note: When cost is closely linked to price, sharing cost data can be considered to be the same as price sharing).
- Do not ask competitors for sensitive data or cause the benchmarking partner to feel they must provide such data to keep the process going.
- Do not ask competitors for data outside the agreed scope of the study.
- Consider using an experienced and reputable third party to assemble and 'blind' competitive data.
- Any information obtained from a benchmarking partner should be treated as you would treat any internal, confidential communication. If 'confidential' or 'proprietary' material is to be exchanged, then a specific agreement should be executed to indicate the content of the material that needs to be protected, the duration of the period of protection, the conditions for permitting access to the material, and the specific handling requirements that are necessary for that material.

4.5 Summary

Benchmarking and competition law are not mutually exclusive. They are both geared to the promotion of quality, innovation, value added and, above all, competitive enhancements. Competition law protects the interest of all through principles of fairness and benchmarking provides opportunity for all through principles of learning through sharing. Words like confidentiality, secrecy, sensitivity can be effectively dealt with through a structured approach and through abiding with and subscribing to the European Benchmarking Code of Conduct.

5 The value of industrial visits

A World of facts lie outside and beyond the World of words

Thomas Huxley

Figures never lie

English proverb

When written in Chinese, the word 'crisis' is composed of two characters – one represents danger, and the other represents opportunity

John F. Kennedy

5.1 Introduction

There is somewhat of a paradox when one wants to examine the real benefit of industrial tours. There is often quoted reference to the fact that industrial visits are analogous to 'industrial tourism' with high cost and no real value. On the other hand, managers need a lot of persuasion and convincing when it comes to the introduction of change. They insist on seeing, since 'seeing is believing'. Industrial visits can be extremely beneficial if there is a final process to manage them and capturing the outcomes in a meaningful way. This is the best way to deal with this predicament, to ensure that the cost of visiting other organizations is well justified and that real outcomes can result from the various visits.

The whole process of making industrial tours worthwhile will be described through a case study, reflecting the experience of Elida Fabergé Ltd. Prior to that, however, the key steps to be observed are covered in the following section.

5.2 The process of conducting industrial visits

It is essential for any organization contemplating visits to other companies to go through the following steps:

1. **Identifying the real purpose of visits:** This is the most critical step in the whole process of conducting visits. Companies need to clearly highlight

the objectives of the visits and the desired outcomes. This will create focus, give structure and lead to positive action. From the point of view of the host organization, it provides the opportunity of preparing, pulling in the right people and getting the information ready for the visitors.

2. **Selecting team members for the visits:** Once again, in order to avoid the criticism of 'industrial tourism', the people involved in the visit must be those who are concerned with the key objectives and the implementation of change once best practice has been identified. By allocating roles and responsibilities, the visits can then be managed very effectively.
3. **Selecting a mission leader:** This is another critical step. Individuals visiting other organizations are wearing their 'corporate hats' and need to act as good 'ambassadors' for their organization. More importantly, they have to be in a position to answer any questions in relation to their business operations and be able to provide a strategic and complete perspective. The team leader must be someone with a senior position.
4. **Preparing the presentation:** It is advisable that a proper presentation of the organization is prepared, using video and other means. The host organization will be interested to learn about the visitors and their organization. The process of benchmarking has to reflect mutuality, reciprocity and the development of proper ongoing links.
5. **Selecting partners:** There are various ways and means for choosing potential partners. Essentially the approach has to be based on:

   - What is the area of interest?
   - What are the key objectives of the visit?
   - What are the expected outcomes?

   The likely list of potential hosts will be based on:

   - Organizations with world-class practices
   - Organizations whose *core* business is reflected by the areas sought to be examined
   - Winners of quality awards and accolades
   - Most-admired companies
   - Existing contacts.

   There has to be a categorizing of the potential list based on:

   - Most-preferred list
   - Desirable list
   - Convenient list
   - No-go list.

   The criteria to be used for the prioritization process should include:

   - Ease of access

- Quality and value of learning
- Geographical location
- Time convenience
- Cost.

Normally one would expect a useful list not to exceed five or six companies. The potential hosts have to be contacted once the list is drawn up, clearly stating the aims and objectives of the visit and enclosing background information. A very important note is that the visits can be reciprocated and the same level of sharing can take place. Once the replies are received, a decision could be made on whether to approach alternatives in the event that there are rejections.

6 **Designing the learning logbook:** There has to be a process for asking all the relevant questions, and also capturing the learning. This vehicle is critical since action plans can only be put together, based on the level of learning achieved.

7 **The briefing process:** All the team members need to be properly briefed on the host organization, and background information needs to be prepared and circulated to all the members. Information could include:

- Organization size, turnover, etc.
- Type of industry, product range
- Business imperatives
- Quality initiatives, innovation.

8 **The debriefing process:** Using the session on questions and answers and the various logbooks, a comprehensive list of points, issues and action can be put together at the end of the visit. Preferably individual roles and responsibilities can be identified in order not to lose momentum and to ensure that the project is going to result in the effective implementation of change.

9 **Closing the loop: feedback to host:** The host partners have expectations. It is therefore important to write back to them, thanking them but also giving them feedback on:

- Most impressive parts of the visits
- Likely issues that could have arisen
- Offers of assistance, help and advice in specific areas.

An open invitation also needs to be included, to reflect the spirit of reciprocity and mutuality.

10 **The role of honest brokers:** In some instances visits can be better managed through the involvement of a third party, such as consultants. They can:

- Facilitate contacts through their large networks

- Help to design the brief for the project
- Work closely with other consultants representing the host partners
- Facilitate the briefing and the debriefing processes.

5.3 The Elida Fabergé Ltd experience

Elida Fabergé Ltd is a wholly owned subsidiary of Unilever plc and is the largest health and beauty products manufacturer in the UK. Sales turnover in 1995 was £98 million at NPS (the price the manufacturer receives), which represents an 11.3 per cent increase on 1994. Of total sales, deodorants and female body sprays represent 26 per cent, hair products 22 per cent, skin care 11 per cent, oral care 7 per cent, and 34 per cent is accounted for by male toiletries. Elida Fabergé Ltd employs 760 people, of whom 390 are directly involved in manufacturing or production activities.

Elida Fabergé Ltd sells over 30 branded products and has significant shares in four core markets:

1	Deodorants/female body sprays	Aerosols, sticks, roll-ons
2	Male toiletries	After-shaves, after-shave gels, body sprays, shower gels, sticks and roll-ons
3	Hair	Shampoos, conditioners, hairsprays, colorants
4	Oral care	Toothpastes and toothbrushes

5.3.1 Company purpose

> Our purpose is to create, produce and sell brands which help people care for their personal health, confidence and appearance and so enhance their sense of well-being. Our aim is to delight all customers with every product and service we offer.

Elida Fabergé Ltd has recently restructured to adopt a process-based approach and to become more integrated and streamlined to give it a regional/global focus. Its major customer base is to the large retail multiples such as Asda, Boots, Safeway, Sainsbury, Superdrug and Tesco. Elida Fabergé Ltd is a 'people' organization and very much admired for its best practices in relation to employee involvement, development and welfare as evidenced by the following statement:

> As a learning organization, we encourage shared responsibility for personal development and training programmes which meet individual and business needs. Through this, we seek to empower our employees as the basis of our growth and success.

5.3.2 Purpose of the study tours

To help Elida Fabergé's efforts to intensify its quality management efforts through the implementation of self-assessment using the European Business Excellence Model, many managers needed to be exposed to this model and convinced of its importance to Elida Fabergé. The aims and objectives of the visits were therefore:

- To study the application of self-assessment for business excellence
- To benchmark the Elida Fabergé Ltd approach against other models
- To learn about benefits and the strategic importance of self-assessment
- To document how self-assessment can drive a TQ culture.

5.3.3 Planning the visit

In all, fifty-four middle managers initially showed interest. Based on expressed interest, an initial list of five potential companies was drawn together and four were eventually visited:

- D2D
- IBM (UK) Ltd
- Britvic Soft Drinks Ltd
- Milliken Industrial Ltd.

5.3.4 Managing the visits

Well before the visits took place, background information on the host companies was put together and circulated among the managers concerned (Appendix 5.1). Following agreement for the visits to take place, scheduled programmes were put together and agreed between the parties concerned. A programme for briefing the teams was also included prior to conducting each visit (Appendix 5.2). Part of the briefing process was to present together a logbook for capturing the key learning outcomes from each visit and to enable each individual to enter their own action plans (Appendix 5.3).

The debriefing process for each visit took place for the purpose of developing an action plan and to highlight key learnings and any issues which may have arisen. In all, a comprehensive range of useful learning points did emerge from the visits and these formed the basis of various action plans. Appendix 5.4 illustrates some of the major outcomes arising from the visits.

Feedback was given to all the hosts as part of the whole process, together with indications on possible future areas of exchange in order to keep the momentum of benchmarking going.

5.4 Summary

The process of industrial visits is extremely beneficial and can add value to reformulating strategy and putting together organizational plans that could have corporate-wide implications. The major benefits of industrial tours are various:

- 'Seeing is believing' and this is the best approach to persuade managers on the value of some change programmes
- A good way for preventing complacency and demonstrating what can be done and achieved
- Prevent 'reinventing the wheel' by quickly exposing managers to best practice
- A good approach for extending managers' learning and development.

Appendix 5.1: Background information on the host partners

Visit to D2D

1. Company background information

- Organized into six separate divisions
- Electronics manufacturing services
- Over 2000 employees worldwide
- Five manufacturing locations
- Sales office in San José, USA and Tokyo in Japan
- 1994 revenue $US450 million
- Wholly owned ICL subsidiary
- In 1989 excess capacity was sold to Sunmicro Systems, a long-standing customer
- ICL split itself in 1992 into twenty-six different businesses which operate autonomously. D2D changed its mode of operation from a cost centre into a profit-making centre
- D2D is in the business of subcontracting technology deals with some direct competitors of ICL as well
- D2D uses a permanent core workforce and also temporary staff
- A wholly owned subsidiary of ICL which is owned (85 per cent) by Fugitsu.

Mission

> To become Europe's leading electronics manufacturing services company, in our chosen markets, by providing products and services worldwide that exceed our customer expectations.

- D2D tries to be a full turnkey provider (a complete service) which includes:
  - Technical and engineering services
  - PCB fabrication
  - PCB assembly
  - System assembly and test
  - Supply, software and documentation services
  - Repair, refurbishment and recycling.
- Five sites occupying 40 acres.

Strategy

- Complete electronics manufacturing solutions
- Ten to fifteen strategic partners
- Leading and emerging companies:
  - IT
  - Telecomms
  - Instrumentation.

Customer interface

Dedicated account management.

- Electronic Data Interchange:
  - E-mail
  - CAD data
  - Schedules and orders
  - Quality feedback
  - Customized performance score cards
  - Easy to do business with.
- Effective competitiveness as defined by the customer is based on *price* and *total cost of ownership*. What drives D2D is to provide all the above to deal with pressures from formidable competitors such as Solectron who are also winners of MBNQA.
- Two and a half years ago Compaq started to lose market share and their strategy was sourcing from outside by buying cheaper components and assembling their computers to differentiate their products through price. They realized that technology and its functionality is the same. This made them realize how important it is to drive costs down.

2. *Quality processes*

- Strategic business model
- Self-assessment and continuous improvement
- Zero defects
- Customer care philosophy
- Quality the D2D way.

The quality road started in 1984 when ICL was at the brink of extinction, saved by the British Government who came in with a £250 million package. This period also coincided with the arrival of new blood, from Texas Instruments, for instance, and their style of leadership was better communications, wider involvement, etc.

Harmonization in working conditions – single-status canteens – favourable working conditions – focus on the business had to be right – making ICL a market-driven business as opposed to a tradition of being engineering/product focused (there was a tradition of over-engineering).

- Top management team published and communicated the ‘ICL WAY’, through vision, communication of the strategy to all employees.
- Used Phil Crosby approach to launch quality (1985). The whole organization were put through formal training (26 000 employees). ICL has gone through a 10-year revolution programme.
- The establishment of the customer–supplier concept was a big challenge.

There is still a lot to do to bring people in the chain, particularly in the service area (such as finance).

- Found that there was at some stage (1990–1991) overemphasis on the internal customer and 'measuring to death' internally, which meant that there was a loss of focus on the external customer. This meant that there has to be a new approach with customer care, strategic quality and deciding where the next steps are.
- This led to the consideration of self-assessment. Initially MBNQA was considered, and then the decision was made to adopt the EQA model. The EQA model had the results section and this was compatible with D2D's strategic expectations of driving with quality.
- The approach used widely now is the strategic business model, based on the EQA model. D2D found that this is a complete business model.
- To get everyone on board and commit process owners at the senior level, each senior manager was given a task to write a draft submission on their key areas of enablers/results and subject it to impartial assessment. This was a revelation in the sense that people started to realize that there is a lot of scope for improvement and perceived performance is well below what is revealed by the assessment exercise.

3. Quality achievements

- 1993 EFQM – 2nd prize winners
- 1993 *Microsun Systems* – 'Outstanding Performance Award'
- 1994 Management Today and Cranfield School of Management 'Best Factory Award'
- 1994 Winner of EQA.

Non-ICL business is going to grow to over 50 per cent.
Actual turnover 1994 – £450 million.
Projected turnover 1996 – £600 million.

4. Services:
A. Engineering services

- Design Process Consultancy
  - Impartial advice on CAE
  - Evaluation of best CAE tools
- Component simulation
- PCB development
- Test engineering
- Product development.

B. Conformance services

- EMC
- Product safety
- Environmental testing
- Telecomms approval.

C. Bare board manufacturing

- Advanced interconnect
  - High-density PCBs
  - MCMs
  - Advanced materials.
- Up to 32 layers.

D. Taking the strain

- Proving new manufacturing technologies
- Process development and control
- Design for manufacturing and test
- Inventory management
- Responsiveness
- Flexibility
- Economies of scale.

5. D2D: the four phases of quality

Conformance quality	– BS 5750–BS 7750 – National Vocational Qualifications – SPC
Customer-driven quality	– Customer Care – Customer Scorecards – prevention Toolset (QFD, DFM)
Market-driven quality	– Best Practice Benchmarking – EQA
Strategic quality	– Strategic Quality Model

QFD is a powerful tool but it requires mature customers to be involved to make its impact more effective.

- Employees who control the process are given various pieces of information including prices of components, margins per unit. This enables them to work out how much rework needs to be carried out if there is a need, in order to break even and maintain the margin.
- The continuous improvement process was started back in 1983 for the following reasons:
  - Worldwide capacity
  - Low margins
  - Technology (geometry shrinks, more functionality, means boards are

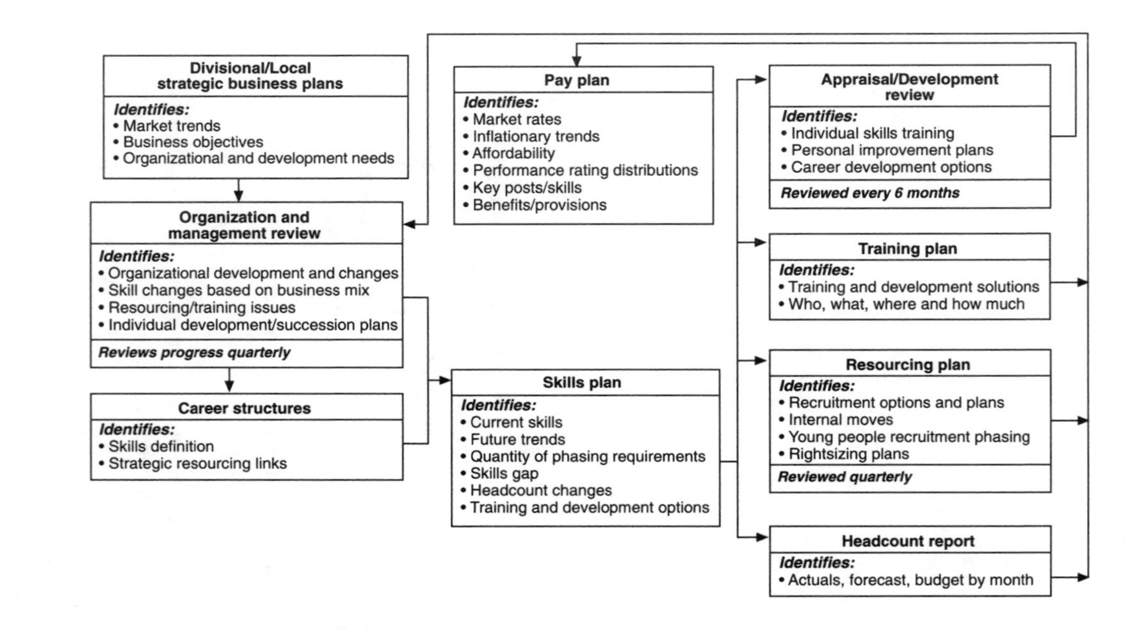

Figure A5.1 *Investing in people – human resource process*

shrinking and this leads to less requirements for manufacturing large quantities)
- Mainframes to PCs
- Customer requirements.

The above factors are drivers and the following list are the things that have been done to deal with the drivers:

Investing in people

As Figure A5.1 illustrates D2D uses a closed loop Human Resource Development model, which intends to deliver the following attributes:

- Harmonization
- Career structures
- Healthcare
- Problem solving
- Opinion survey
- Recognition
- Participation
- Process ownership
- National Vocational Qualification (NVQ)

thus leading to:

Responsibility and ownership

- Thirty per cent of employees have been trained in the use of quality tools and techniques such as FMEA, SPC, etc.
- Recognition takes many forms, from a gift voucher to the ICL Excellence Scheme, where there are Gold, Bronze Awards . . . The Delta innovation scheme, to encourage ideas, people who enter at least two or three ideas per month can have their names put in a draw and they could win a prize.
- There are two sets of objectives: personal and team objectives and they are measured separately.

Education and training

- In 1986 introduced Crosby
- In 1988 ICL introduced its own approach called Quality Improvement for the Individual (QIFTI)
- Plus, over years:
  - QIFTI for managers
  - QIFTI refresher
  - QIFTI for temporary employees
- Prevention tools:
  - SPC

- Customer care
- Tools and techniques
  - Problem solving
  - Corrective action
  - Teamwork
- National Vocational Qualifications (NVQ)
- Training is made bespoke for specific requirements and to suit the skills of the people who operate specific processes.

6. Performance measurement

COST OF NON-CONFORMANCE (cost of quality = cost of conformance + cost of non-conformance. CONC is measured and compared with percentage of revenue.

- D2D measures hard aspects such as manufacturing operations and soft operations such as personnel, safety, finance. For example, for employee attrition, cost of training is measured.

COST OF QUALITY ⇒ MEASUREMENT OF COST ⇒ PRIORITIZED OPPORTUNITIES ⇒ CORRECTIVE ACTION ⇒ MAXIMIZED PROFITABILITY

Customer scorecard:

Measurement criteria: hard and soft measures:

- Achievement to forecast
- Flexibility and responsiveness to service request
- Ease of doing business
- Delivered quality
- Failure analysis and corrective actions
- Cost competitiveness
- Technical support.

Each criterion is weighted: for example, delivered quality is 30 out of 100.

- At each point of interface, people introduced their own scorecard. *The customer owns the scorecard* (for example, throughout the supply chain). The results are compiled every month to review performance and also to review overall targets set.

EQA Model

D2D considers this to be the missing link between continuous improvement and business results.

- Why was EQA used?
  - To establish a model against which process in quality management can be objectively measured
  - To enable the delivery of profitable growth through the measurement architecture. STRATEGY ⇒ PROCESSES ⇒ CUSTOMER SATISFACTION ⇒ BUSINESS RESULTS

Visit to Britvic Soft Drinks Ltd

The company

The starting point for Britvic was in the nineteenth century when the British Vitamin Products Company started producing flavoured mineral waters in Chelmsford, Essex. Today, Britvic is owned by Bass, Allied Domecq and Whitbread which have a joint 90 per cent share, the remaining 10 per cent being held by Pepsi-Cola International. Bass is, however, the major shareholder.

Britvic has a turnover which exceeds £400 million and supplies 20 per cent of all carbonated soft drinks consumed in the UK. Britvic's corporate mission is: 'to provide the best brands, the best service and the best value in a way which meets both our customers' and our consumers' needs.'

Britvic is committed to Total Quality Management and to dealing with environmental challenges in its quest of becoming a world-class competitor.

Britvic is part of an FMCG highly competitive and growing market. In the UK alone, more than 8 billion litres of soft drinks are sold, and this gives a market larger than wines and spirits, milk, tea and coffee. By the end of the century, this is expected to rise to 10 billion litres.

Britvic sells 950 million litres of ready-to-drink soft drinks annually in nearly 400 different flavours, shapes and sizes. They produce eighteen leading brands and supply more than 250 000 retailers. The company has, over the years, produced very successful advertising, with investment of up to £40 million per year. Some of the most famous ads include:

- 'Lipsmackinthirstquenchin'(Pepsi)
- 'Been there, done that' (Pepsi Max)
- 'You know when you've been Tango'd' (Tango).

Effective competitiveness for Britvic is assisted by a big distribution centre (National Distribution Centre) based in Leicestershire, which opened in 1994. This centre can handle up to 70 per cent of the demand (700 truck-loads of soft drinks per day).

Famous brands

Tango is the UK's number-one fruit carbonate and third best-selling soft drink. It was launched in 1994. Britvic juice is the product for which Britvic is best known, sold in pubs. Other brands include Liptonice (first carbonated ice tea) launched in 1994 by Britvic Soft Drinks, Pepsi and Van den Bergh Foods; Pepsi was invented in 1898 and sold in 149 countries around the world; R. Whites is a clear lemonade made with real lemon juice; 7UP is the leading lemon and lime carbonate brand and manages to outsell its nearest competitor by 5:1.

The operation

- More than half of its employees work in operations, managing and controlling the supply chain.
- Over £1 million is spent daily on raw materials, goods and services.
- There are six factories throughout the UK, each specializing in specific products of particular types of packaging.
- Company operates on Just-in-Time basis, working closely with all its suppliers. Some materials are sourced from around the world.
- Quality Assurance procedures are in place to monitor production lines which operate at a rate of 1500 cans per minute.
- All production and distribution sites have ISO 9000 accreditation.
- Company has heavily invested in new technology and people skills.
- Employee reward systems at Britvic are linked to personal performance.
- Britvic operates on a more or less flat organizational structure. Its training budget annually exceeds £1 million.

Total quality management at Britvic

The mission of Britvic was launched as the **3Bs** (Best Brands, Best Service and Best Value). Called 'Business Excellence' the quality programme at Britvic was led by their managing director (Stephen Davies). He meant TQM to mean a complete change in behaviours and attitudes and the starting point, therefore, was to seek opinions and views and to generate corporate commitment:

> 'I want to make the company a byword for quality, service and excellence in our industry – the Supplier of Choice – because customers can rely upon us to deliver that little bit beyond their expectations. We shall become a company that aspires to match world class performers in terms of Brands, Service and Value and overall business.'

Stephen Davies and his senior management team took a roadshow to all employees, selling them the concept of business excellence and the importance of linking quality to bottom-line results.

Business Excellence was not launched as a matter of urgency, since the company in 1994, for instance, generated profits which were up by $5 million. It was to inject a visionary approach to strengthen their competitiveness and follow the footsteps of model organizations such as Rank Xerox and Rover Group. Britvic wanted to learn from the best. In order to tap into the various opportunities, Britvic recognizes that its culture must:

- Create a common understanding of problem-solving techniques and provide training to enable Britvic to improve its way of working.
- Encourage employees to generate ideas and implement the vast majority of them.
- Form functional teams and working groups to maximize progress.
- Understand what customers will want five to ten years from now, and how to exceed their expectations to achieve total customer commitment.
- Work together to achieve quality, cost efficiency and profits.

At Britvic it is strongly believed that people involvement will make the difference in succeeding with business excellence and changing the culture positively. There are already many initiatives which have been linked to the business excellence programme, such as:

- People involvement
- Investors in people
- Quality management system
- Leadership training.

The Business Excellence Model of Self-Assessment:

Britvic chose Rank Xerox Business Services Ltd (part of Rank Xerox Ltd) for introducing self-assessment. The Business Excellence Certification (BEC) Model was developed from The Malcolm Baldrige, EQA and Deming Prize Models. It is represented by six key elements. The categories represent:

- Management leadership
- Human resource management
- Business process management
- Customer and market focus
- Quality support and tools
- Business results.

(See Figure A5.8.)

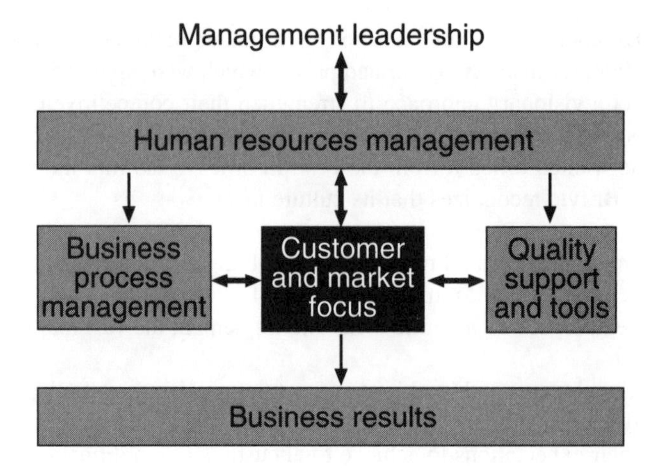

Figure A5.2 *The Business Excellence Model used by Britvic Soft Drinks Ltd*

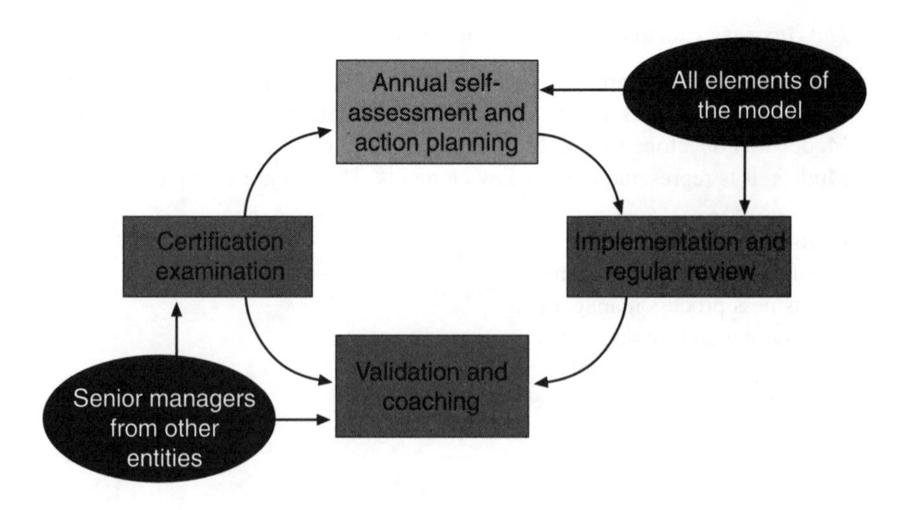

Figure A5.3 *Self-assessment process used at Britvic Soft Drinks Ltd*

The rating scale

All items must be rated based on the information obtained. A rating of 4 is the threshold for a sound system and is based upon the three interrelated dimensions of **results**, **approach** and **pervasiveness**. The diagram below provides refinement and precision in gauging performance and improvement potential.

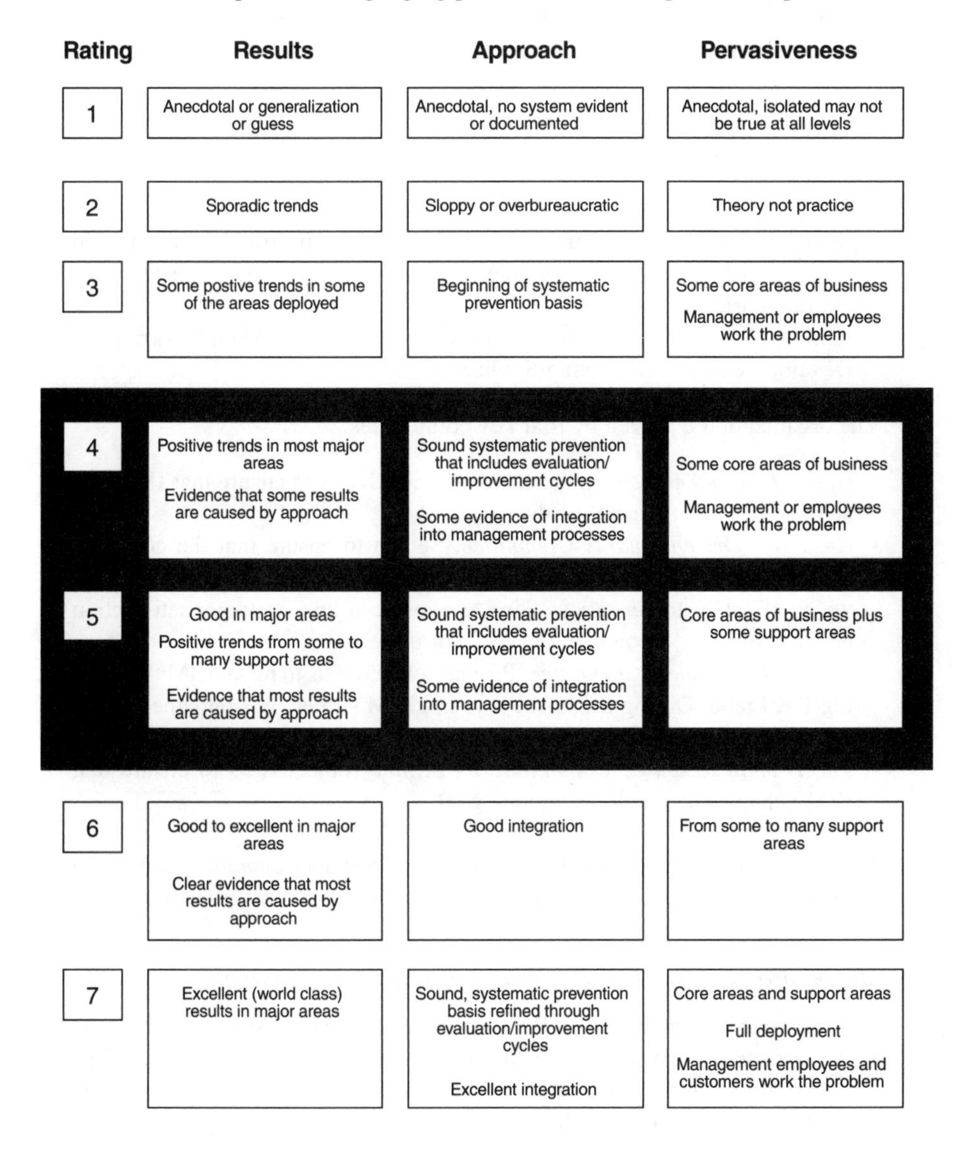

Figure A5.4 *Scoring process for self-assessment within Britvic Soft Drinks Ltd*

Visit to Britvic Soft Drinks Ltd

The self-assessment framework within Rank Xerox is applied in all the business units annually, in a systematized manner, and the assessment exercise is conducted by third-party senior managers from sister sites. The process itself is described in Figure A5.3. As far as the rating scale is concerned, the scoring is done on a scale of 1–7, and conducted from the point of view of the approach, deployment and results. (Figure 5.4).

Visit to IBM UK LTD

IBM UK Ltd:

- IBM UK Ltd is a subsidiary of IBM Corporation. It provides information technology hardware, software solutions and services to customers throughout the world.
- IBM UK's operations include Product Divisions (Manufacturing & Development) and Marketing Services.

The organization is driven by four key committees:

- *Group Investment Committee*: Its main objective is to ensure that IBM UK achieves its two financial goals.
- *Customer Driven Quality Committee*: Set up to ensure that the company achieves its delighted customer goal. It meets monthly to review projects which affect customer satisfaction. As a result of this, customer satisfaction has improved by 8 points in the space of three years.
- *Brand Management Committee*: Primary objective is to make IBM the leading UK brand. During the last few years, IBM's brand has remained in the top five most recognized brands in the UK.
- *The Human Resource Committee*: Its primary objective is to ensure that IBM achieves its employee morale goal.

In addition to these four committees, there is a *Strategy Committee* whose role is to identify markets which are likely to emerge in the next seven years.

Market driven quality (MDQ): IBM's approach to self-assessment

MDQ is based on the Malcolm Baldrige National Quality Award criteria. The principles are exactly the same, to enable self-assessment of all of IBM's business units. The driving principle behind MDQ is the need to focus on customer satisfaction through a total quality management approach. The MDQ assessment criteria are built upon the following core values and concepts:

- Market-driven quality principles
- Leadership
- Continuous improvement and innovation
- Full involvement and empowerment
- Cycle-time reduction
- Design quality and problem prevention
- Long-range outlook
- Management by fact
- Partnership development
- Public responsibility.

MDQ is, therefore, used by IBM as a strategic tool for enhancing its competitiveness. The way the company has decided to launch it is perhaps best described in the following statement:

> MDQ is directed toward gaining market share, profitability and retaining satisfied customers. It demands constant sensitivity to emerging customer and market requirements, and measurement of the factors that drive customer satisfaction. It also demands awareness of developments in technology, and rapid and flexible response to customer and market requirements. Defect and error reduction and elimination of causes of dissatisfaction contribute significantly to the customers' view of quality and are thus also important parts of Market-driven quality. In addition, the ability to prevent defects and errors as well as to respond quickly to defect and error correction is crucial to improving both quality and relationships with customers.

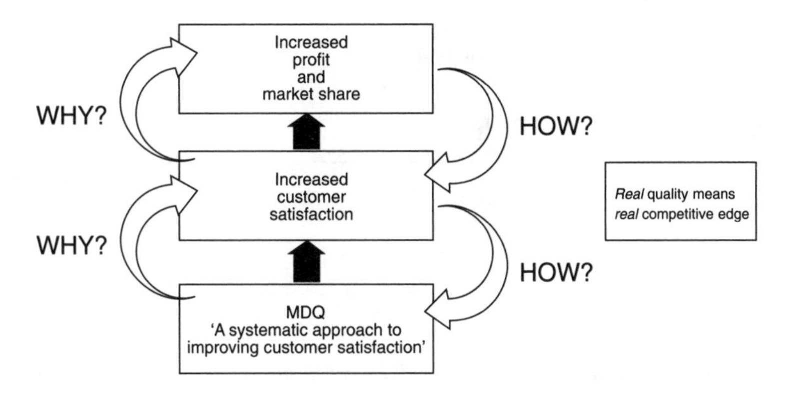

Figure A5.5 *MDQ and its impact on customer satisfaction and market share*

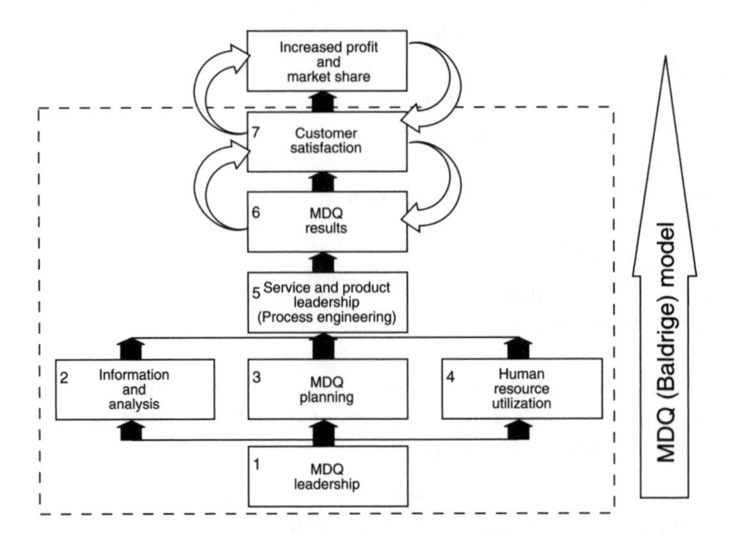

Figure A5.6 *The strategic management process using MDQ at IBM*

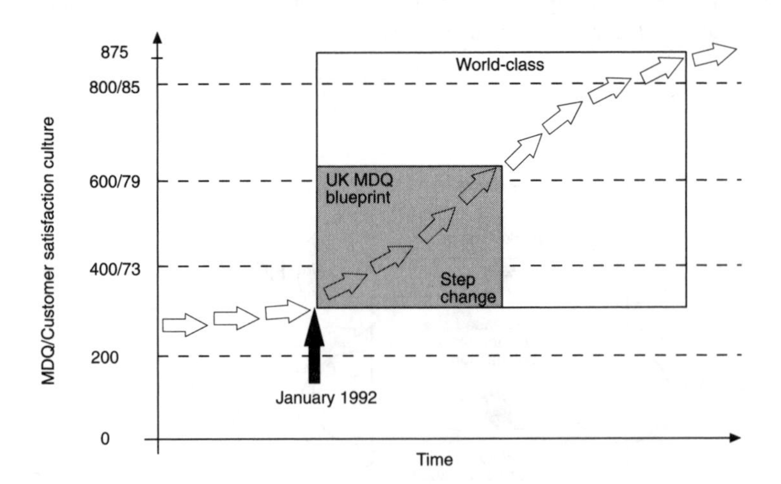

Figure A5.7 *Using self-assessment to build a corporate culture of quality at IBM*

	Categories and Items	Points	
1.0	**Leadership**		**90**
1.1	Senior executive leadership	45	
1.2	Management for quality	25	
1.3	Public responsibility	20	
2.0	**Information and analysis**		**80**
2.1	Scope and management of quality and performance data and information	15	
2.2	Competitive comparisons and benchmarks	25	
2.3	Analysis and uses of company-level data	40	
3.0	**Strategic quality planning**		**60**
3.1	Strategic quality and company performance planning process	35	
3.2	Quality and performance plans	25	
4.0	**Human resource development and management**		**150**
4.1	Human resource management	20	
4.2	Employee involvement	40	
4.3	Employee education and training	40	
4.4	Employee performance and recognition	25	
4.5	Employee well-being and morale	25	
5.0	**Management of process quality**		**140**
5.1	Design and introduction of quality products and services	40	
5.2	Process management – product and service production and delivery processes	35	
5.3	Process management – business processes and support services	30	
5.4	Supplier quality	20	
5.5	Quality assessment	15	
6.0	**Quality and operational results**		**180**
6.1	Product and service quality results	75	
6.2	Company operational results	45	
6.3	Business process and support service results	25	
6.4	Supplier quality results	35	
7.0	**Customer focus and satisfaction**		**300**
7.1	Customer relationship management	65	
7.2	Commitment to customers	15	
7.3	Customer satisfaction determination	35	
7.4	Customer satisfaction results	75	
7.5	Customer satisfaction comparison	75	
7.6	Future requirements and expectations of customers	35	
Total points			**1000**

Figure A5.8 *1992 IBM MDQ assessment categories and items*

Approach	Deployment	Results	
Fully effective prevention-based system of continuous improvement; excellent integration across the business	Fully deployed and used in all areas of the business and all support areas	World-class results sustained in all business and support areas; clear competitive advantage; results obviously caused by the approach	100%
System well developed and tested; excellent integration in most areas	Fully deployed and used in all major areas of the business and most support areas	Excellent results in most areas of business and many support areas; sustained, improving competitive advantage; clear evidence that results are caused by the approach	80%
Well planned and documented with evidence of refinement through evaluation/improvement cycles; good integration in many areas	Deployed and used in all major areas and many support areas	Good results in major areas of the business and positive trends in some support areas; much evidence they are caused by the approach	60%
Sound, prevention-based system including evaluation/improvement cycles; some evidence of integration	Deployed and used in most major areas and some support (peripheral) areas	Positive trends in most major areas with some evidence they are caused by the approach	50%
Beginning of prevention-based system; methods, tools, techniques defined; integration begun in some aspects of the business	Deployed and used in many major areas of the business	Positive trends in key areas of the business	40%
Reactive systems; awareness of need for prevention-based systems and integration across business	Beginning in some parts of the business	Some results in areas where deployed	20%
No system or integration	Not deployed or used	No evidence that results are caused by the approach	0%

Figure A5.9 *Assessment scoring system at IBM*

Approach	Deployment	Results
'Approach' refers to the methods used to achieve the purposes addressed in the items. The scoring criteria used to evaluate approaches include one or more of the following, as appropriate: • The appropriateness of the methods, tools and techniques to the requirements • The effectiveness of the use of methods, tools and techniques • The degree to which the approach is systematic, integrated and consistently applied • The degree to which the approach embodies effective evaluation and improvement cycles • The degree to which the approach is based upon quantitative information that is objective and reliable • The degree to which the approach is prevention-based • The indicators of unique and innovative approaches, including significant and effective new adaptations of tools and techniques used in other applications or types of business • The degree to which the organization provides evidence that a process exists when the Area or Item asks 'how . . .?'	'Deployment' refers to the extent to which the approaches are applied to all relevant areas and activities addressed and implied in the items. The scoring criteria used to evaluate deployment include one or more of the following, as appropriate: • The appropriate and effective application by all parts of the organization to all processes and activities • The appropriate and effective application to all product and service features • The appropriate and effective application to all transactions and interactions with customers, suppliers of goods and services, and the public	'Results' refers to outcomes and effects in achieving the purposes addressed and implied in the items. Trends should include enough data points (3–5 years recommended) to demonstrate a sustained rate of improvement over time. The scoring criteria used to evaluate results include one or more of the following: • The quality and performance levels demonstrated and their importance • The rate of quality and performance improvement • The breadth of quality and performance improvement • The demonstration of sustained improvement • The comparison with industry and world leaders • The organization's ability to show that improvements derive from its quality practices and actions

Figure A5.10 *Scoring guidelines for MDQ at IBM*

Scoring System

The system for scoring items is based upon three evaluation dimensions: (1) approach; (2) deployment; and (3) results. All items require organizations to furnish information relating to one or more of these dimensions as outlined in Figure 2. Specific criteria associated with the evaluation dimensions are described below. Scoring guidelines are outlined in Figure 3. The results of applying the scoring weights shown in Figure 3 to the maximum point values shown in Figure 4 determine the assessed value. Figure 4 also reflects the summary of the item level point values to determine the total assessed value.

Visit to Milliken Industrial Ltd

Milliken Industrial Ltd

> Milliken and Company is dedicated to the continuous improvement of all products and services through the total involvement of all associates. All associates are committed to the development and strengthening of partnerships with our external customers and suppliers.
>
> We will continually strive to provide innovative, better quality products and services to enhance our customers' continued long-term profitable growth by understanding and exceeding their requirements and anticipating their future expectations.
>
> Roger Milliken
> Chairman and CEO, Milliken

- Founded in 1865 by Seth Milliken and William Deering in Portland, Maine, this company was initially involved in trading cotton and cotton goods. William Deering later moved to Chicago to form Deering Harvester, today Navistar.
- Roger Milliken took over as CEO in 1947, and under his stewardship the company has become probably the largest privately owned corporation in the USA and is totally self-financed.
- One of the world's leading textile companies and also one of the largest, with its headquarters established in Spartanburg, South Carolina. There are sixty-three manufacturing plants and 14 000 associates worldwide, producing more than 48 000 different textile and chemical products.
- In addition to an extensive contract carpet business based at Wigan, worldwide Milliken produce fabrics and yarns for protective and fashion wear, car seats, tennis ball covers, printer ribbons, mats, towels and chemicals and dyes for a range of uses associated with textile and other industries.*
- In 1964, Milliken set up its first European manufacturing plant in Bury,

*Lloyd, B.J. (1993) *Meeting the customers' real needs*, MBA Dissertation, Bradford University Management Centre.

Lancashire and subsequently also in Wigan. Since then, additional plant and sales offices have been established at Ghent in Belgium, Risel and St Julien in France and Morke in Denmark. The European division has grown dramatically since 1970, achieving a turnover of £40 million in 1992.

- **Milliken at Wigan:** The manufacture of modular carpets commenced at Beech Hill, Wigan, in 1986. The facility manufactures carpets for the European market serving commercial and industrial markets. The plant has installed carpets through its extensive dealer network, in prestigious locations throughout Europe, including Lloyd's of London and many commercial and industrial headquarters of international businesses.
- Much of the technology used in the Milliken plants has been designed and developed by Milliken Research Corporation.
- The standard range of carpets includes over 1000 designs and colourways. Milliken's CAD facility allows the creation of unique designs to customer requirements.
- **Milliken's management system:** This is illustrated in Figure A5.11 and reflects that the achievement of the Milliken goal of Profitable Growth, using a Total Customer Satisfaction approach, by focusing on **Q**UALITY, **C**OST, **D**ELIVERY, **I**NNOVATION, **S**AFETY, **M**ORALE and **E**NVIRONMENT (**QCD IS ME**). The service element of the system was added on in later years in line with the original model by Ichiri Miyauchi.
- Milliken have measured their costs of quality in relation to their manufacturing processes since 1986. Figure A5.12 illustrates the improvement in the management of non-conformances throughout the past few years.
- Milliken uses a process of continuous improvement involving individuals in all parts of the organization. The programme is known as *Opportunities For Improvement (OFI)*. Figure A5.13 illustrates the development of OFI and its growth over the years.
- Figure A5.14 illustrates the buying structure where the interface with the customer comes through a complex route. Distribution is, on the whole, through independent deals, contractors who operate on behalf of the customer.

Milliken Industrial Ltd: the TQM journey

Milliken's headquarters are in Spartanburg, South Carolina. It employs over 12 000 people referred to as 'associates'. It has 28 businesses which produce 48 000 different textile and chemical products, ranging from apparel and automotive fabrics to speciality chemicals and floor coverings. Its annual sales are in excess of $1bn.

Total Quality at Milliken

Milliken embarked on introducing TQM more than 10 years ago and is achieving great results under the banner of continuous improvement. In 1981

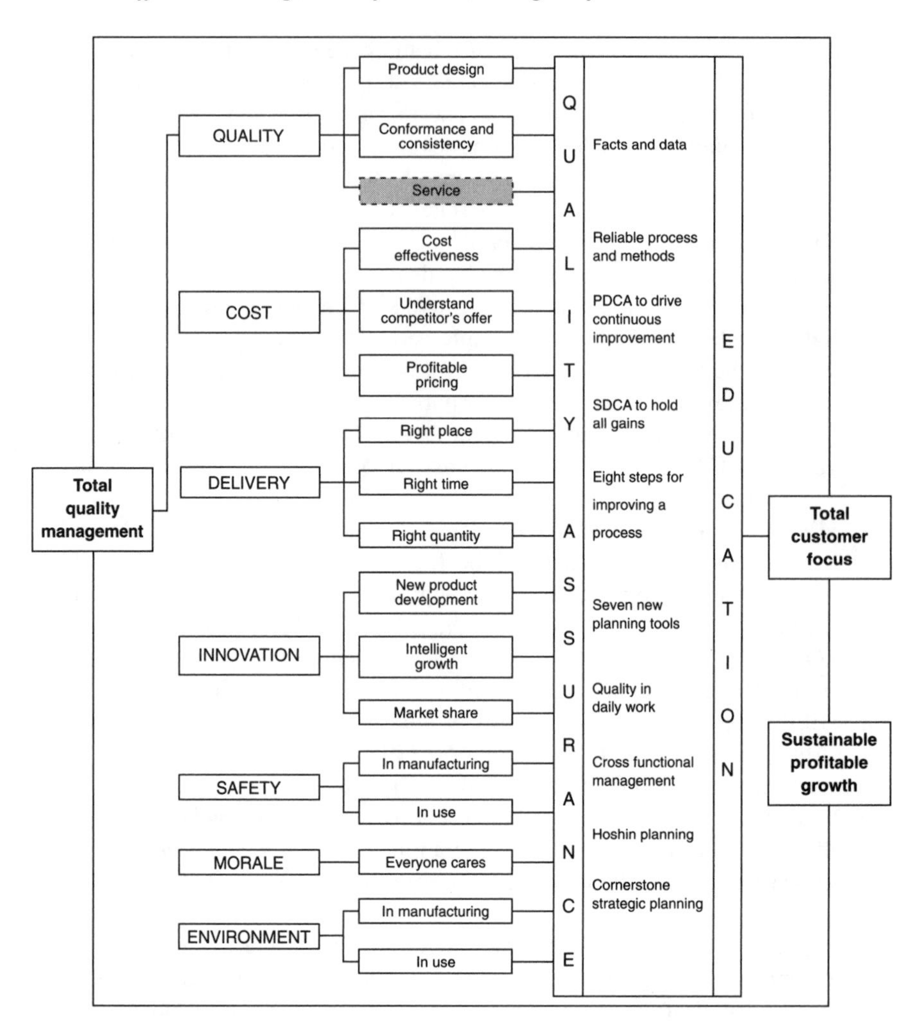

Figure A5.11 *Milliken's management system*

Milliken introduced a TQ programme called 'pursuit of excellence' (POE) based on a commitment to providing total customer satisfaction. The company's goal (mission) is: 'to provide the best quality products, customer response, and service in the world, through constant improvement and innovation with a bias for action.'

The commitment is to eliminate poor quality or what is referred to as 'off-quality' and to improve, continuously, quality standards for the benefit of the end customer, improving the quality of 'first' quality.

The deployment of TQM at Milliken takes place through heavy involvement of people (associates), as stated by their quality policy.

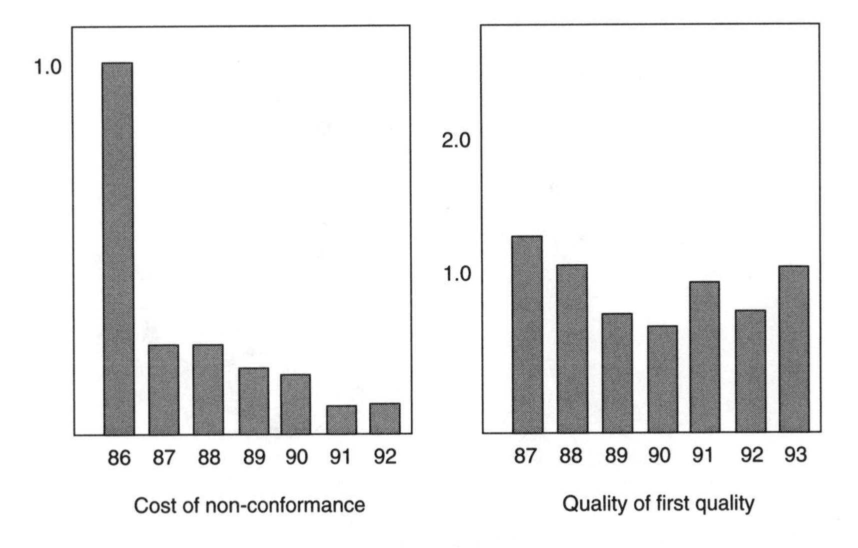

Figure A5.12 *The management of quality costing at Milliken*

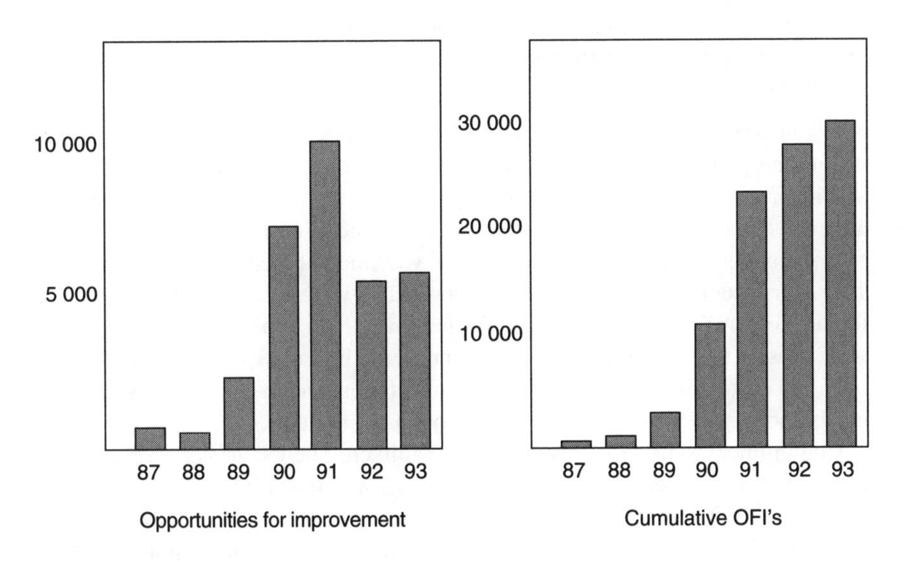

Figure A5.13 *Opportunities for improvement (OFI)*

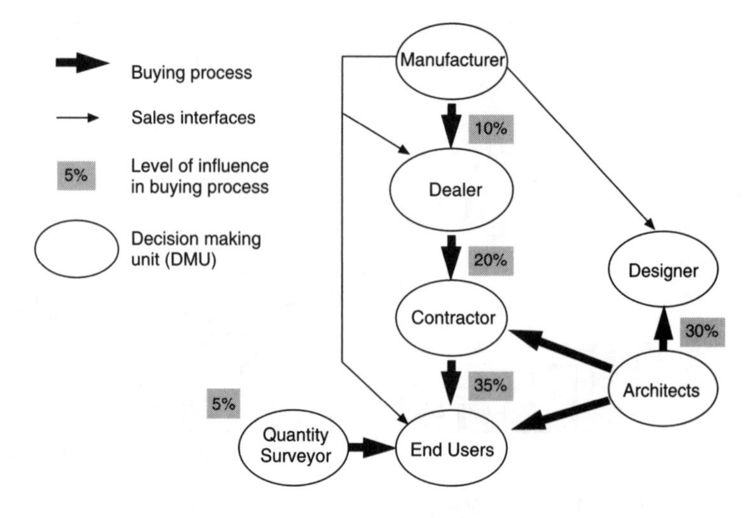

Figure A5.14 *Buying structure at Milliken Industrial Ltd*

> Milliken & Company is dedicated to the continuous improvement of all products and services through the total involvement of all associates.
>
> All associates are committed to the development and strengthening of partnerships with our external and internal customers and suppliers.
>
> We will continually strive to provide innovative, better and better quality products and services to enhance our customers' continued long-term profitable growth by understanding and exceeding their requirements and anticipating their future expectations.

Milliken uses a policy committee and quality council to deploy its quality vision and goals. It has a flat management structure, where people work in self-managed teams enjoying a high degree of autonomy and freedom. People are even empowered to stop production lines if there is a need.

Continuous improvement at Milliken is very much based on team work. In 1988, for instance, 1600 corrective action teams were formed to look at different problems. The company also formed over 200 supplier action teams to work closely with Milliken's suppliers in order to improve quality levels. Well over 500 customer action teams were formed to respond to customers' various needs and requirements and also to iron out any problems.

Commitment to its people is evident at Milliken. There is heavy investment in training and in recognizing people's efforts and rewarding them for their contributions. Training is also extended to the company's customers and suppliers. While Milliken has been very successful in increasing its customer base considerably through its commitment to quality, the number of suppliers, on the other hand, was reduced considerably and since 1981 the company has reduced its supplier base by 70 per cent.

The TQM approach used at Milliken is based on Tom Peters and Nancy Austin's book *A Passion for Excellence*. Its main components are:

- Care of customers
- Constant innovation
- People
- Leadership (managing by wandering around).

This approach basically starts with a good understanding of customer requirements, a capability of exploiting market opportunities and being highly innovative and optimizing the internal resources such as people and strong leadership to be constantly proactive and ahead of competition.

Performance measurement at Milliken

Milliken takes the measurement area very seriously. This is acknowledged by Roger Milliken, who, when asked about what has been learnt along the way to winning the Baldrige Award, replied:

> We have learnt the importance of measurement. If you can't measure it, it isn't worth doing. You must be able to look at the result of your efforts and then compare them to a scoreboard.

Because of the 'what gets measured gets improved' philosophy, Milliken has been recognized as a leading company by various surveys. It was found to be number one in its field based on fifteen customer-satisfaction measures.

The driving force for achieving high results is quality improvements which are capable of achieving customer satisfaction which, ultimately, have a direct impact on company business performance. Milliken uses a set of measures for this.

The 'ten-four' programme which was introduced in 1988 is aimed at improving key areas tenfold during a four-year period. The measures used are:

1 *Cost of non-conformance*
2 *Quality of first quality* – physical properties, product consistency and aesthetics
3 *Cycle time and throughput time*
4 *Customer lead time*
   (a) All sample lead times
   (b) Production lead times
   (c) 100 per cent on-time delivery
5 *Customer responsiveness* – both internal and external
6 *Total customer satisfaction*

7 *Innovation*
 (a) New products
 (b) Differentiated products
8 *Safety*

Safety, for example, is one of the most important measures and this is why the company has developed their safety policy:

> The safety and health of all its people is of primary importance to Milliken & Company. Milliken will devote resources to train its people to perform their jobs safely to ensure equipment can be operated in a safe manner, to eliminate workplace hazards and to comply with applicable safety and health laws and regulations. Milliken believes that all injuries are preventable, all health risks are controllable and management is accountable.

The company also does benchmarking on all its products and services against a list of over 400 competitors. The information collected is used to assess its performance and also to identify opportunities in the marketplace. As a result of benchmarking, Milliken realized that it was lagging behind its competitors in delivery performance. This created the required motivation to improve its record and, as a result, Milliken improved its delivery performance to 75 per cent in 1984. By 1988, on-time delivery became 99 per cent, which is a best-in-class performance.

Achievements – future challenges

Milliken has achieved in various aspects under the banners of TQM. The following are only some of the examples:

- It has managed to reduce its management positions by 700 since 1981 and increased cross-functional, self-managed and autonomous work groups, called 'process improvement' specialists.
- There has been a 77 per cent increase in the ratio of production to management associates.
- Since 1984, over 7500 visitors received training at Milliken's dedicated facilities each year.
- Most manufacturing processes are monitored by real-time monitoring systems for error detection and elimination.
- Since 1981, a 60 per cent reduction in the cost of non-conformance has been achieved (reductions were in areas such as discounts for poor quality, payment for freight on customer returns, etc.).
- A huge database was created providing information on environmental and safety issues, suppliers and customers' issues.
- Advanced technology was introduced to increase customer responsiveness and give customers access to information, e.g. on design parameters. This has heavily impacted on cycle time for new product development.

Milliken enjoyed various recognitions and received over forty major customer quality awards. The biggest prize Milliken won is perhaps the Baldrige Award. This is providing the company with the motivation to move forward with its quest for achieving superior performance. When asked about what will be the next challenge since winning the MBNQA, Roger Milliken replied:

> For two years in a row, we've conducted our own internal Baldrige Quality Competition because we believe continually challenging ourselves to understand and meet these demanding criteria is crucial. . . . it's fun to swim downstream instead of upstream – to talk with customers and suppliers as partners, not adversaries.
>
> Also we must learn to highly value the customer and get to know him better. We must learn to delight the customer.

Appendix 5.2: Samples of agendas for visits and pre-visit meetings for formal briefing

Visit to Britvic Soft Drinks: Briefing Process

6.00–6.30	Arrival and coffee	
6.30–6.40	Welcome and importance of self-assessment to Elida Fabergé	HG
6.40–7.00	Current status within Elida Fabergé	PF
7.00–8.00	Business excellence certification (the approach used within Britvic)	PL
8.00–8.30	Learning logbook	MZ
8.30–9.00	Creating the baseline for benchmarking	(syndicate work)
9.00	Dinner	

Britvic Soft Drinks/Elida Fabergé Ltd

Meeting on Tuesday 12 December
at the Britvic National Distribution Centre,
Magna Park, Lutterworth

1.	Assemble/coffee/introductions		30 mins	9.00–9.30
2.	Opening session	SJD	10 mins	9.30–9.40
3.	Elida Fabergé presentation	HG	30 mins	9.40–10.10
4.	Britvic presentation	SJD	30 mins	10.10–10.40
5.	*Coffee break*		20 mins	10.40–11.00
6.	Questions/answers on self-assessment	All	90 mins	11.00–12.30
6.	*Lunch*		60 mins	12.30–13.30
8.	Tour of NDC	All	90 mins	13.30–15.00
9.	Re-convene/coffee/final Q&A session	All	25 mins	15.00–15.25
10.	Next steps	MH	15 mins	15.25–15.40
11.	Evaluation	All	15 mins	15.40–15.55
12.	Close	SJD	5 mins	15.55–16.00

Benchmarking Study Visit ICL (D2D) – Stoke on Trent

Agenda

7 March

18.00–18.30	Arrival and coffee at the Manor House Hotel	
18.30–18.50	Current status of self-assessment in Elida Fabergé	(M. Zairi)
18.50–19.50	Business excellence certification	(M. Zairi)
19.50–20.30	Learning logbook	(M. Zairi)
20.30–21.00	Creating the baseline for benchmarking	(Syndicate)
21.00	Dinner with Dayvon Goodsell, Group Quality Manager, ICL	(D2D)

8 March

09.00–09.15	Arrival and coffee at ICL (D2D)	
09.15–09.45	Elida Fabergé presentation	(R. Firth)
09.45–10.30	ICL (D2D) company presentation	(D. Goodsell)
10.30–11.00	*Coffee*	
11.00–12.00	Presentation on quality journey (Process management, achievements, future challenges . . .)	
12.00–13.00	Guided tour	
13.00–14.00	*Break for lunch*	
14.00–15.00	ICL (D2D) Presentation on benchmarking applications and the implementation of Best Practice	(D. Goodsell)
15.00–15.30	The role of self-assessment in building business excellence within ICL (D2D)	
15.30–15.45	*Coffee*	
15.45–16.30	General discussion on self-assessment, implementation issues, benefits, strategic aspects. . .	
16.30–17.00	Feedback on key learning from visit (Elida Fabergé personnel only)	
17.00	Depart	

Appendix 5.3: Logbook for capturing individual learning

Benchmarking Self-Assessment:
Application and Benefits

Participants' Visits Logbook
Company:

Enablers

Type of process	Good practices	Issues arising
1 Leadership		
2 People management		
3 Policy and strategy		
4 Resources		
5 Processes		

Softer issues/Remarks

Results

Type of area	Good measurement practices and outcomes	Issues arising
1 People satisfaction		
2 Customer satisfaction		
3 Impact on society		
4 Business results		

Softer issues/Remarks

Scoring summary sheet

Company: Visiting:

1	Enablers criteria									
Criterion number	**1**	**%**	**2**	**%**	**3**	**%**	**4**	**%**	**5**	**%**
Criterion part	1a		2a		3a		4a		5a	
Criterion part	1b		2b		3b		4b		5b	
Criterion part	1c		2c		3c		4c		5c	
Criterion part	1d		2d		3d		4d		5d	
Criterion part	1e		2e		3e		4e		5e	
Criterion part	1f									
Sum of parts										
Divider		÷6		÷5		÷5		÷4		÷5
Score awarded										

Note: The score awarded is the arithmetic average of the % scores to the criterion parts. If applicants present convincing reasons why one or more parts are not relevant to their organization it is valid to calculate the average on the number of criteria addressed. To avoid confusion (with a zero score) parts of the criteria accepted as not relevant should be entered 'NR' in the table above.

2	Results criteria															
Criterion number	6	%			7	%			8	%			9	%		
Criterion part	6a		×0.75		7a		×0.75		8a		×0.25		9a		×0.50	
Criterion part	6b		×0.25		7b		×0.25		8b		×0.75		9b		×0.50	
Score awarded																
Aggregate average																

3. Calculation of total points			
Criterion	Score awarded	Factor	Points awarded
1. Leadership		× 1.0	
2. Policy and strategy		× 0.8	
3. People management		× 0.9	
4. Resources		× 0.9	
5. Processes		× 1.4	
6. Customer satisfaction		× 2.0	
7. People satisfaction		× 0.9	
8. Impact on society		× 0.6	
9. Business Results		× 1.5	
Total points awarded			

- Enter the score awarded to each criterion (of both sections 1 and 2 above).
- Multiply each score by the appropriate factor to give points awarded.
- Add points awarded to each criterion to give total points awarded for application.

Score justification

Learning outcomes against objectives

Key learning objectives	Outcomes	F = fulfilled PF = partially fulfilled UF = unfulfilled
1		
2		
3		
4		
5		
6		
7		

Post-visit action plan

Area of concern	Specifications
•	⇒
•	⇒
•	⇒
•	⇒
•	⇒
•	⇒
•	⇒
•	⇒
•	⇒
•	⇒
•	⇒
•	⇒
•	⇒
•	⇒
•	⇒
•	⇒
•	⇒
•	⇒
•	⇒
•	⇒
•	⇒
•	⇒
•	⇒
•	⇒
•	⇒
•	⇒
•	⇒
•	⇒
•	⇒
•	⇒
•	⇒
•	⇒
•	⇒
•	⇒
•	⇒
•	⇒
•	⇒
•	⇒
•	⇒
•	⇒
•	⇒

Appendix 5.4: Examples of some of the outcomes distilled from individuals' logbooks

Visit to D2D: Distillation from Individual Logbooks

Leadership

- Vision – 'Best in Europe'
- Clear view from directors
- Communication of business plan to **all** shopfloor employees
- Being prepared to implement vision
- Excellence in communication
- Senior people do the training
- MBWA
- Vision + Charisma of CEO – **drivers**
- Champions within the business process (business model)
- Open/honest

People management

- Move people across different functions
- Flexibility of temporary labour
- Talk to people constantly
- Up to **4** surveys annually
- Elimination of status (e.g. all overalls are the same)
- People move horizontally as well as vertically
- People contracts, temps with 1-year contracts, TPM, investment in training
- Career streams
- Training not part of personnel
- Training programme fits in with business plan and tailored to customer needs

Policy and strategy

- HR strategy has very good links with business plan
- Design for Manufacture – key process for success in this industry
- Always check payback – even environmental decisions
- Customer focus – methods for feedback (scorecard)
- Organized in business units (COMs)
- Benchmarking – comparison with similar sized businesses – finding the real competitor
- Effective use of assets (e.g. let out extra building space)
- True customer organization

- Clearly defined niche
- Manufacturing excellence
- Partner for chosen customers (large)
- Growth means change – change means mistakes (but mistakes mean learning)
- Very clear direction 2-year plan ⇒ 4-year plan

Resources

- Actively recruit from competitors
- Consultants to benchmark all aspects of the business
- Split into cells
- Multi-skilling of better people + engineers
- Close scrutiny + understanding of resources (i.e. cost, utilization, etc.)
- All in the operations
- Facilities management
- Temps used to achieve flexibility (1-year contracts)

Processes

- Organized in COMs but leaders in expertise formally share this across the units
- Yield is predicted and cost is modelled
- Complexity of product design
- The business model has owners
- Customer focus
- Flexibility + responsiveness give the edge
- Policy and strategy setting benchmarked
- Comprehensive multi-faceted quality process – conformance, customer,
- Market-Strategic (4 phases)
- Benchmark everything!
- 24-hour production
- Capital base
- Efficiency
- Training run by team leaders

People satisfaction

- Firm and fun leadership
- Teams with clear objectives
- Team (not individual) bonus
- Annual opinion survey (move to 4 a year)
- Feedback + action from surveys is key
- Emphasis on response rate too
- Recognition scheme

Customer satisfaction

- Growth!
- Accreditation
- Quality Awards (best contractor by Sun Microsystems)
- Customer does the scoring/evaluation for D2D
- Always results in both sides coming away with changed views of customer requirements
- Benchmarking partnership with customers
- Cornerstone of business is quality: many systems inc. ISO 9001, 14001, TPM, QFD, etc.
- Don't measure customer satisfaction. Customer does

Impact on society

- Commitment to improve water quality through reduction of copper effluent
- ICL has plant which reclaims old components
- High emphasis on environment (makes business sense)
- BS 7750 – among first 6 companies
- ISO 14001

Business results

- Growth
- New customers
- Responsiveness = key to success + competitive edge (i.e. willingness + speed)
- All measurements look at: 1 – actual; 2 – target; 3 – benchmarks
- 30% growth in 1995; 50% growth estimate for 1996
- Diversified business
- Aggressive growth plans

Soft issues

- Cross-team competition happens – senior management ensure this does not create barriers across teams
- Change is accepted as a way of life – adaptability is a key skill to long-term survival
- Risk taking in situations of 'do nothing and you'll die – change and you might die' (if you get it wrong), but you might survive and grow
- Benchmarking/self-assessment, quality awards, etc. are all added fertilizers to the growth prospects
- Managers are trained in (+ do) self-assessment. Company believes managers can't assume their roles properly unless they can apply self-assessment

- The method of applying self-assessment most preferred is focus groups (questionnaires give least value)
- Profitability comes from leadership
- Improvement plans built in business plans. Everything has to have pay-back, e.g. marketing collateral required from ISO 9000/ISO 14001, etc.
- Helping customers/partners achieve ISO 14001 (passing on as added value/knowledge)
- We need to decide who to benchmark each process, how far have we got?
- Managers are trained as assessors against the business model. They led self-assessment and assess other companies. They also run training courses
- New meaning to TPM (total productive management): getting away from machine efficiency to A × P × Y (availability × performance × yield)

Visit to Milliken Industrial Ltd: Distillations for Individual Learning Logbooks

Leadership

- High visibility of notices/values
- Roger Milliken's effect on people
- In pursuit of excellence is a process and not a programme
- Passion and commitment, dedication of leader and personal involvement
- Way beyond cascading
- Roger Milliken is 'one of us'. He attends training sessions with other managers
- People will take empowerment when they feel safe to do so
- Change is led from the very top
- Regular meetings to ensure clear understanding

People management

- Gives people a structured way of improvement
- New methods are cascaded and taught by senior management (not just handed over for implementation. Line managers have to show their understanding and commitment)
- No bonuses are paid for quality improvement activities
- 100% response rate to annual attitude survey
- Everyone treated the same
- Importance of people is conveyed throughout the history of Milliken
- Time given to training
- No door/open office
- Increased visibility/management by walking about
- Grading people not the jobs (commitment to multi-skilling)
- Goal clarity and communication (*Hoshin* Planning)

- Work with people and customer
- Partnership policy (associates)
- Drive out fear
- Self-esteem makes people feel important
- Management training – minimum of 40 hours per annum

Policy and strategy

- Ability to anticipate customer requirements and then exceeding them
- Quality improvement is everyone's job, not a separate department
- Policy development and goal cascade
- Partnerships for profit
- Road to excellence is the vision (avoid quality word)
- Profit with people not through them
- Clarity of business position in market
- Recognition that defects can occur in all business areas
- Continual change
- Raging impatience with inaction
- Measurement of customer satisfaction
- 80% of time spent on 'creating the environment'
- Very clear policy and well communicated down the line
- Clear measures, 15 years of experience with quality. Evidence of its workability is through the winning of prestigious awards
- Top management involvement in search for improvement and the finding of new ways of managing and learning of new methods
- Emphasis on quality and cost because quality and on time delivery are top issues for customers

Resources

- Heavy investment in research facilities and manufacturing technology. Presently leading the field in many areas of technological advantage
- Design process well resourced and customer friendly
- No piecework pay
- No commission, profit sharing or bonuses
- Pay in advance for potential. Keep good employees
- Innovative equipment design, manufacturing facilities, IT systems
- No quality department. ISO owners. No debts
- 15 000 people/45 plants, 130 years old/leadership
- Big research centre
- Managers' time is to create and sustain culture
- Time allocation

Processes

- Positive philosophy – process is the problem not the people
- Measuring things which are only important to the customer
- Error cause removal
- Keep scores on display with no blame
- Annual surveys of customer satisfaction (by phone) and improvement actions set up for following year
- Based on associates and clear ownership
- OFIs
- Recognition and sharing
- Worldwide leadership for quality, cost, delivery, innovation, safety, morale, environment
- Good systems for capturing ideas
- Good use of technology
- The measurement of customer satisfaction is more important than measuring profits

People satisfaction

- Stopped paying for output or sales. Pay for quality and customer satisfaction
- No commission, no bonus, no profit sharing
- Drive out fear
- Keeping scores displayed
- 90% associates submit OFIs. Actions are followed up
- £1 voucher given for each OFI
- Recognition of 100% attendance
- Noticeboard recognition of good performance. Motivate don't threaten
- Change the system otherwise people get bored
- No blame – people will respond
- Moved from 'workers are the problem' to 'managers are the problem'
- Car park space/thank you cards
- Start taking decisions/empowerment

Customer satisfaction

- Customer action teams
- Use of unstructured telephone customer surveys
- Focus on Milliken's performance against leading competitor not others
- Survey outcome influences Business Managers' salaries
- Customer satisfaction trends are more important than profits through measuring managers' current year performance
- Working through partnerships
- Displays of customers' needs/wants

- Ask and measure
- Defect free in everything not doing it right first time
- Product quality and on-time delivery always top 2 requirements by customers
- Customer satisfaction is a predictor of forthcoming business performance
- Presentation of customer complaints, faults analysis
- Data gathered on quality, on time, handling lates, price, attitude (top 5 indicators)
- Quality is first thing mentioned by associates
- Direct links between customers and factory over quality

Business results

- Defect levels reduced by 61% since 1981
- Pay suppliers promptly leads to reduced prices and better attention/supply
- Continuous growth
- High quality – excellent performance
- Baldrige/EFQM award

Soft issues/Remarks

- Decisions are all taken down from Roger Milliken – this impacts in equal ways on team management/consensus decision making
- Insanity is doing what you always do and expect different results
- Unless 60% of time is spent creating and sustaining the environment/culture – most methods learnt from consultants are useless
- People turnover worldwide is 2.5% (very low)
- It is impossible to give empowerment – people will take empowerment when they feel ready
- Employee attitudes develop aptitude
- No blame – display results
- Don't need kicking – people respond to scores
- Drive out fear (don't shoot the messenger)
- At Milliken – no commission, no bonuses, no piecework, no profit-sharing
- Still on fire, ready, aim organization (thriving on chaos)
- Raging impatience with inaction + flexibility of mind – a do it organization
- Plant tour guide with safety hazard warnings and plant identification
- Vocabulary (employee [associate/self-esteem], staff [support], manager [leader], problem [opportunity])
- Use thank you cards
- Culture change: single status, no offices, management visibility
- Safety management systems devised by associates

Impact on Society – *was not an area covered during the presentation or the tour itself.*

Acknowledgements

The author is greatly indebted to the following managers who made this project a total success:

Peter Fairclough (Corporate TQ Manager, Elida Fabergé Ltd)
Malcolm Holden (Quality Manager, Britvic Soft Drinks Ltd)
Dayvon Goodsell (Quality Manager, D2D Ltd)
Roger Emms (Customer Services Director, Milliken Industrial Ltd)
Tim Claxton and Phil Hanson (IBM Consulting Group)
Paul Leonard (Rank Xerox Quality Services Ltd).

6 The value of benchmarking awards

The majority of businessmen are incapable of original thought because they are unable to escape from the tyranny of reason

Ogilvy

Athletes and companies improve their performance in similar ways – by repeatedly working towards higher goals

Nakamura

The higher the goals, the better the management

Geneen

6.1 The European Best Practice Benchmarking Award: The process for winning with benchmarking

6.1.1 Background to the award

This innovation was launched by the Benchmarking Centre Ltd in the UK at its Second Annual Forum held in November 1994. The objects of the award are to promote and recognize best practice benchmarking and to:

- **Identify** the best examples of benchmarking in Europe
- **Facilitate** the sharing of best practice in benchmarking
- **Share** real-life experiences
- **Dispel** the myths surrounding benchmarking.

6.1.2 The process of entering the award

The process is based on the submission of a case study. The structure and content of the entry should include:

1. A description of the benchmarking process and its constituent steps
2. The implementation of the process
3. The selection of the process to be benchmarked
4. An understanding of the company's own process design and its performance

5 The methods of research for identifying and selecting the benchmarking processes and partners
6 The planning for data collection, visits, interviews and research
7 The data collection procedure, analysis and results
8 The acceptance of changes and their implementation
9 The measurements taken before and after benchmarking
10 The management commitment and the composition of the project team
11 How the technique of benchmarking was integrated into the existing culture and approaches for improvement
12 The identification of the lessons learnt and the achievements from the benchmarking undertaken.

6.1.3 Criteria used for scoring

A selected panel of judges is put together to assess all the entries and the criteria used are presented in Table 6.1. Essentially the judges will use the criteria to determine the best implemented project in the context of business imperatives, how the experience is put together to assist culture development for benchmarking and what benefits may have accrued both in hard and soft senses.

The award is currently in its second year with two winners and two runners-up. In the following sections, both the first winning and runner-up entries are discussed.

6.1.4 Winning with benchmarking: the Hewlett-Packard Finance Ltd story

When Karen Prior-Smith (Business Support Manager) presented at the Second Benchmarking Symposium in the UK she made it very clear that the key trigger of the (IMPACT) benchmarking initiative was HP's Quality maturity system (QMS), illustrated in Figure 6.1. She stated that benchmarking is an integral part of the TQM system which places great emphasis on process management and analysis.

Indeed, because the QMS system starts with the importance of corporate goals and objectives through customer focus, it led to the generation of project IMPACT, particularly since a customer satisfaction survey revealed that there are a lot of improvements in priorities and, in particular, customer service.

6.2 Application of benchmarking at Hewlett-Packard

At Hewlett-Packard benchmarking is defined as 'comparing your business processes to perceived "best-in-class" processes within other organizations in an effort to make significant improvements in performance'. Benchmarking at

Table 6.1 *Criteria of assessment*

Leadership based	**Reasons for selection of benchmarking as a technique (/process)** • Comparison, competitive analysis, Measurement, business process re-engineering • Involvement of customers and suppliers • Continuous improvement ethos **Sponsorship of benchmarking** • Show visible involvement of management • Pertaining to customer satisfaction • External promotion of benchmarking • Training	**Use and commitment of available resource balanced against organization size** • Provision of resources • Position with total quality management philosophy with respect to company size **Evidence of innovation in selection of benchmarking process** • Techniques – lateral thinking, external input, method used • Customers and suppliers input • Geography and industry wide
Process based	**Use of a structured/consistent approach to benchmarking** • Process exists and owned • Consistent team understanding of process • Reviewed and modified **Appropriateness of team composition** • Facilitator, expert • External input (customer/other units) **Demonstration of own process design, performance and enablers' understanding** • Process owned and measured • Techniques used to identify strengths, weaknesses, improvements • Documented • Use of external inputs, e.g, customer surveys	**Comprehensiveness/applicability of background research** • Techniques – journals, databases, lateral thinking • People – consultant, customers, suppliers, academics • Questionnaire – format • Planning **Probity in the interfacing with benchmarking partners abiding by the Code of Conduct** • Code of Conduct – each element • Interface with partners – written, verbal, feedback
Results based	**Process of making it happen** • Analysis of data • Acceptance of key enablers to own process and team • Management support • Implementation plan **Implementation of change and measurement of business improvement** • Changes identified versus changes implemented • Measured effect of changes – process, people, business results • Benefit to customer	**Evidence of win–win relationship with benchmarking partners** • Building a partnership relationship • Quotes! • Lessons learnt

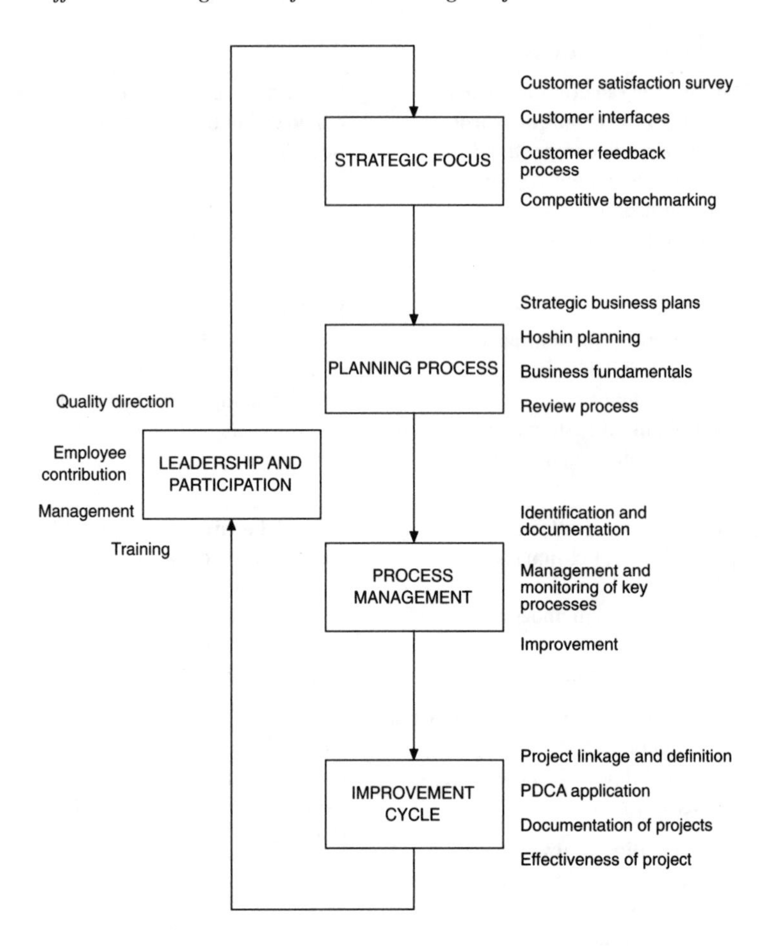

Figure 6.1 *The HP quality methodology: quality maturity systems (QMS)*

Hewlett-Packard is facilitated by the quality department and follows a process developed in the UK over the last five years. The process adheres to Deming's Plan–Do–Check–Act cycle.

Process management is a key responsibility empowered to employees. They have the responsibility to:

- Identify their key processes
- Document their key processes
- Measure the effectiveness of their processes
- Improve their processes.

Employee empowerment is apparent from the managing director's processes throughout the organization down to field engineer and sales rep level.

Currently the UK has around 1000 processes documented and managed, the majority to ISO 9000 standard.

Hewlett-Packard's products have a perceived high quality value among their customers. Its people are highly trained and specifically selected and recruited for their skills. Process management facilitates the linkage between these three key areas of the business. Profit is a financial measure of business success but the people and processes must work together to contribute to the profit (Figure 6.2).

Benchmarking is a management tool used extensively at Hewlett-Packard both internally and externally. It is used to drive continuous improvement through the adoption and adaptation of best practices. Benchmarking is used by Hewlett-Packard 'to provide the external performance perspective needed to motivate entrepreneurial behaviour'. Not only have they experienced performance improvements as a direct result of benchmarking but employees have been motivated and stimulated through its effective deployment. This has occurred as a direct result of increased cross-functional learning and the promotion of teamwork (Figures 6.3 and 6.4).

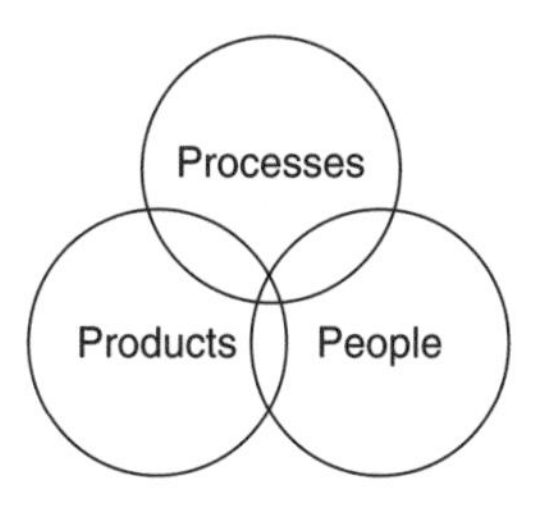

Figure 6.2 *The three Ps of business*

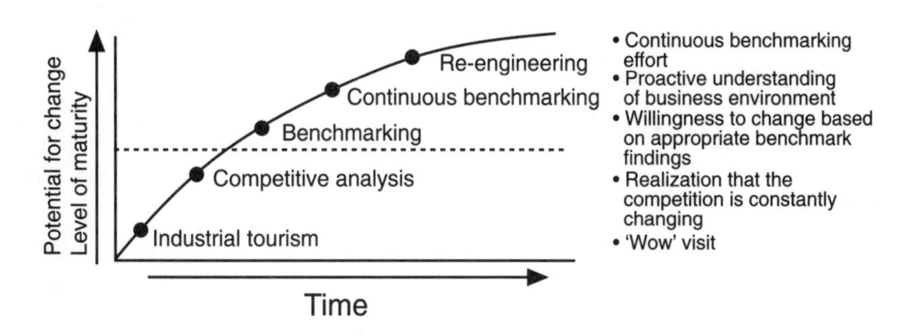

Figure 6.3 *Benchmarking and other improvement tools*

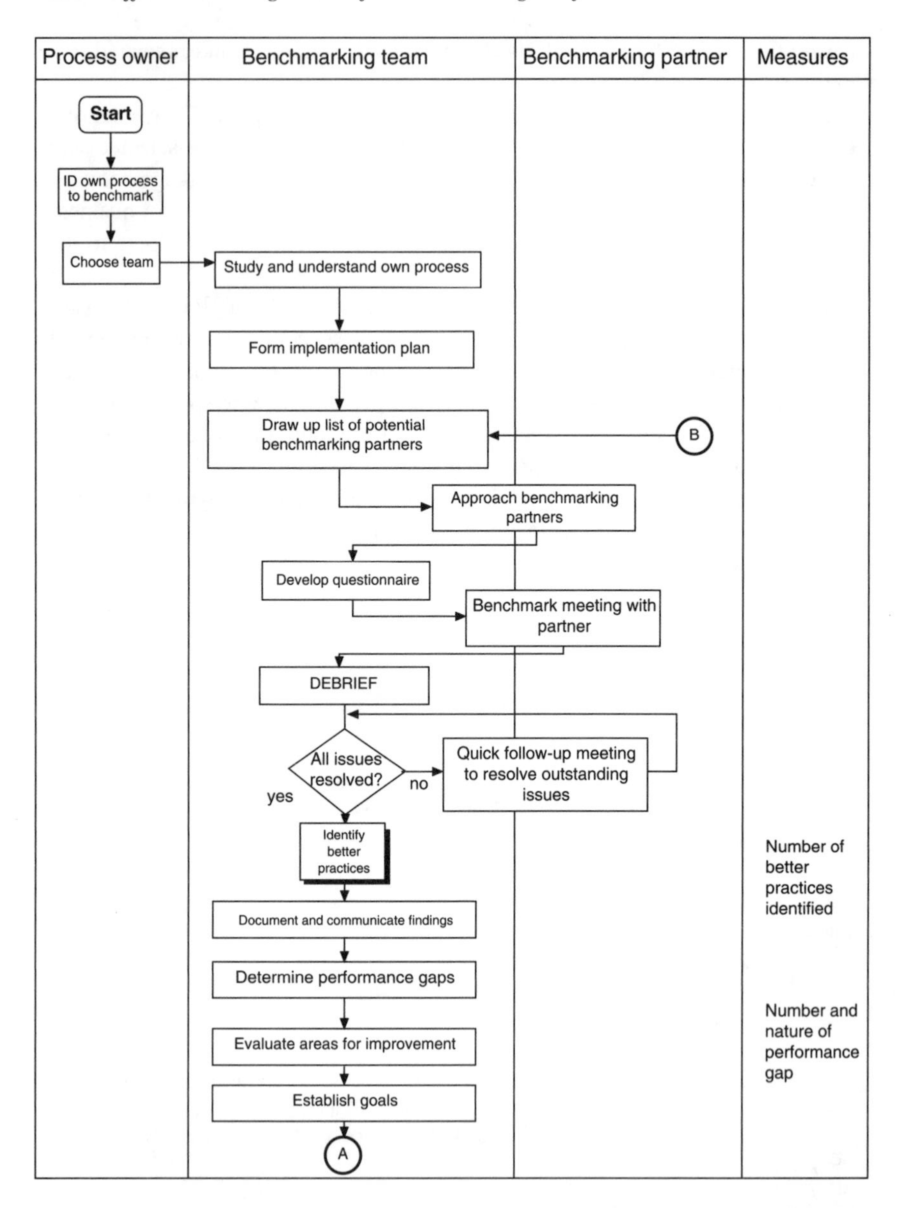

Figure 6.4 *The external process for benchmarking*

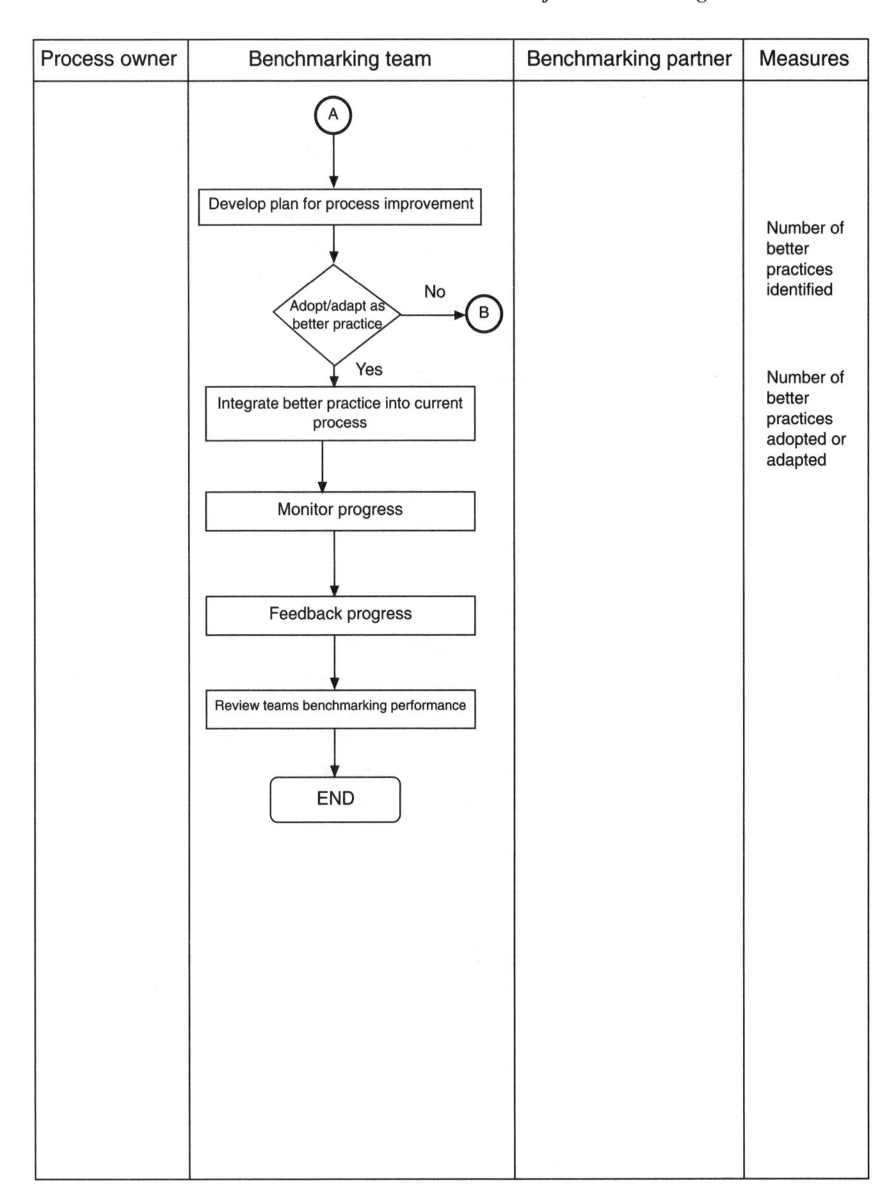

Figure 6.4 *The external process for benchmarking (continued)*

Plan

Team Information

Sponsor: Ian May, Director and Business Unit Manager of Hewlett-Packard Finance and Remarketing
Team: Karen Prior-Smith, Contracts Administration Manager
Paula White, Senior Contracts Administrator
Christine Burke, Senior Contracts Administrator
Facilitator: Yin Wong, Quality Consultant

The reason for starting this benchmark

The Finance & Remarketing Business Plan identified customer service as an area where we could maximize our competitive advantage. Our vision statement was defined as providing consistently reliable and constantly improving service to our customers.

Objectives

- FRD Contract Admin to provide a professional, pleasant friendly service which maximizes our ability to conduct business over the phone through the effective use of our systems and maximizing our productivity.
- To have a clear process for handling any customer complaint which will turn a potentially damaging situation into a positive one.
- To effectively reward and motivate staff who demonstrate a service level beyond the call of duty.

Q2 95

Further visit to Company X by benchmarking team to look at specific areas of customer service, process improvement and employee management. Questions prepared and sent to Company in advance and agenda for day submitted back to us.

Q2 95

Visit undertaken 2 May 1995.

Questions submitted to Company X

Service questions

- How do staff answer the telephone?
- How do you guide a customer through a call and get the correct information from customer?

- Can we listen in to a call?
- How do you measure performance?
- How do you control filing?
- Do you use microfiches? Do you destroy documents?
- What systems do you use?
- How do you use technology to its best effect?
- Can we see some customer documents such as invoices, letters, folders?
- How do you handle customers' objections? Is the customer always right?
- How do you determine if to charge for a service, e.g. sending a copy of a statement?
- How do you divide workload?
- What happens when several people are off sick at one time?
- How are teams' responsibilities divided?
- How do you measure productivity?
- Can we see training in action?
- What incentives do you offer employees?
- How do you measure effectiveness of training?

Organization

- Number of people in operation?
- Their roles/responsibilities?
- Their perception of their own role?
- Which processes they manage?
- Ownership?

Systems

- Use of IT/telecomms infrastructure?
- Handling of paperwork generated received?
- Filing/follow-up processes?

Measures

- How do they measure themselves?
- How do they determine success factors?
- Productivity/volumes per head?

Rewards

- Pay/benefits
- Employee satisfaction
- Conditions/environment/culture

Customers

- Needs versus satisfaction levels
- How are these determined and measured?

General

- Areas where they do not perceive they are good
- How they seek to improve.

Do

Hosts

Call Centre Manager
Customer Service Training Manager
Head of Business Process Management
Customer Services Support Unit Manager
Customer Services Manager
Customer Development Manager
Training Development Manager
Information Analysis Manager
Internal Communication Manager

Agenda

10.30–11.00 am	Welcome and overview of the company
11.00–11.45 am	Induction Course
11.45 am–12.00 pm	Business Process Review
12.00–1.15 pm	Lunch
1.30–2.30 pm	Call Centre Visit
2.30–3.00 pm	People Issues with Training Appraisal, Career Development Recruitment, Induction Course
3.00–3.30 pm	Business Measures & Communication i.e. Business Barometer
3.30–4.00 pm	Conclude

Check

The performance gap is currently considerable with Company X operating superior practices at the time of the visit (Figure 6.5). The following information will be used to narrow the gap.

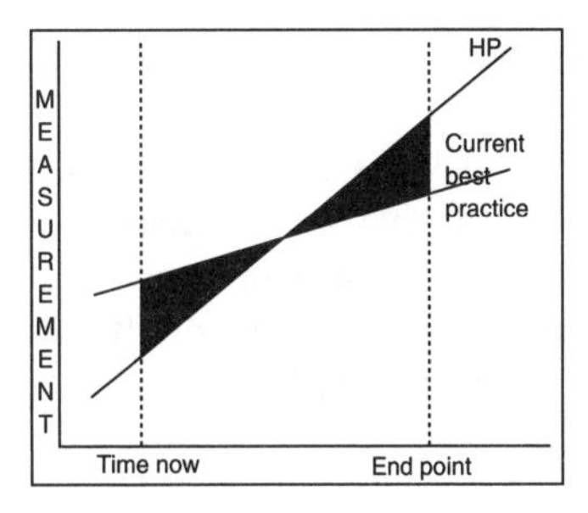

Figure 6.5 *Performance gaps*

Calibrating the gap

Customer focus

Everything is geared towards providing the best service level possible to the customer. Satisfaction figures quoted bear witness to this fact. We were told that little marketing is done. Despite this, 10 000 accounts are set up a week. Most customers are upmarket, professional demanding a high level of satisfaction. More than 50 per cent of acquisitions are via recommendations, 80 per cent of customers are very satisfied and 91 per cent are not interested in returning to other types of accounts.

Training

Seven weeks' induction training of which the first week consists solely of communications training. Their philosophy is to get people to think about communication before they learn product knowledge. They get used to using these skills during remaining product training. They believe their communications training is second to none.

They conduct accreditation at the end of the induction course which consist of role-play type tests which can last up to 2 hours. There is a 70 per cent pass rate at the first attempt. Ninety-five per cent pass within first or second attempts. The second attempt is taken 4 or 5 days after the first and after the team leader has given some coaching/extra training. They also conduct accreditation after other training courses where possible.

During the training the new recruits become accustomed to giving and receiving feedback. They have weekly 1–1s and written reviews. The training uses the system which they will be using when they are doing the job.

Feedback is important and plentiful at all stages of training (and thereafter). Once trainees are on the job, the team coach monitors their calls, ensuring they can cope. The 7 weeks' training is followed by 9 months' probation, accompanied by a development log, which details what they have done in the call centre over the past 9 months. After this period their business and development needs are assessed.

There are various levels of knowledge within the teams, culminating in a 'supergroup'. These are the people whose time is less likely to be taken up with calls all day. They tend to be the people who will follow up on enquiries off-line, taking the pressure from the front end. It was interesting to note that the team leaders are expected to spend 10 per cent of their time on the phones, taking calls.

Approach to business processes improvement

Company X ask themselves the following questions:

- Where can gains be made?
- Do we need to do the process at all?
- Can the existing process cope with increasing volumes?
- Are we planning to grow in that area? And therefore is it worth improving?

The emphasis is on improving efficiency, while at the same time keeping costs down and quality up.

Customer needs are at the top of the hierarchy, services being there to fulfil those needs with processes in place to deliver those services. There are about 150 processes, which need to be mapped out and examined to determine whether they need to be improved or are needed at all.

The following are some aspects of process evaluation

- Cost
- Volume
- Time taken
- Expectations
- Bottleneck
- External opportunities (growth).

Appropriate surveys used to ascertain customer expectations. Surveys are ongoing. Very efficient. System not the most high tech, but fast! Serves their needs. Looks like needs recognized before system developed, and not the other way round. Call logging and information access all on one system, including capability to log custom complaints, 95 per cent of which are dealt with on-line.

For customers requiring more detailed information (e.g. mortgage information) there are specialists to handle. Selling products is not top priority, but is encouraged (and seems to happen automatically, without the banking reps realizing they are selling!). BWs encouraged to 'work off the conversation'. The system prompts BR to follow up on information sent out previously.

Information sharing

Every available space is used to advertise, whether it be figures, measures, information (walls, ceiling and free-standing boards used). Very effective and colourful.

Atmosphere/motivation/fun aspects

There is a great 'buzz' in the call centre. This fact was reiterated by the BRs we spoke to at lunch. They spoke of this as what made their job enjoyable (as well as being part of a great team of people).

Budget set aside for rewards, to be used in whatever manner they wanted. BRs themselves were asked what they would like. Opted for 'gold envelope'. This envelope could contain anything from use of parking space in front of building to an M&S voucher. Reward could be for the smallest achievement, and could be presented spontaneously on the spot to the lucky BR, or more formally.

Friendly competition encouraged between teams (to encourage selling, getting leads, new accounts, etc.). Fun days set up to increase awareness of core values (games, T-shirts, balloons).

Complaints handling

If a customer makes a complaint it is logged on the system using only a couple of key depressions. The banking rep is trained to handle the complaint but will pass the customer to their team leader if necessary and 95 per cent of complaints are handled in this way. Five per cent are passed on to customer complaints department. Complaints are monitored to see if a trend occurs.

There is a Snakes and Ladders 'Do's and Don'ts of Complaint Handling' board. They have a board highlighting customer complaints which have been made with pictures of sheep on it to make it more noticeable. Helps to stop same mistakes being made again.

Sales attitude of banking reps

Part of delivering excellent customer service is offering new services. They record everything of a sales nature, e.g. when they send out literature. This then flashes up on the screen the next time the customer calls so that they can follow up with the customer. 'Just before you go, we are asking customers today if they own a car . . .'

Additional information

Banking rep stays in job for an average 2 years. Team leaders spend 10 per cent of their time on taking calls to keep them in touch. Rest of time spent monitoring, motivating, appraising and developing. Each TL has 12 BRs. Work is not checked but is monitored from time to time sometimes with TL listening in to call. Seventeen calls taken by each BR an hour. Answer the phone 'Hello. This is Company X. How can I help you? I'm just going to take you through your security details.' BR told customer what she was doing, used customer name and said 'Thank you for waiting' after putting customer on hold.

They are in process of formalizing lateral moves as they are seeing a greater demand for career development. Encouraging people to learn broader technical skills. Historically they have promoted people quickly. Problem now is that they have some people in jobs which they should not be in. Have new level of team coach between BR and TL which is a development level.

Act

The following pages show:

- Implementation plans and their current status for training (Table 6.2)
- Complaints handling (Table 6.3)
- Reward and recognition (Table 6.4)
- The new complaints process (Figure 6.6)
- Follow-up letter
- Definition of a complaint
- Complaints handling traffic lights (Figure 6.7)
- Copy of certificate awarded to employee.

Impact of benchmark

- Hewlett-Packard has developed a winning relationship with Company X and hosted a reciprocal visit. The inbound process for benchmarking is shown in Figures 6.8 and 6.9.
- Finance & Remarketing now have a complaints process in place operating to a 24-hour turn-around time with follow-up letter.
- A formalized programme is currently being developed.
- Recognition now exists for work well done.
- There is less rework due to the new process and training.

6.5 Winning with benchmarking: the IBM National Call Management Centre story

When Berit Mortlock (Benchmarking Process Manager) presented at the Second Symposium on Benchmarking in the UK she emphasized that even

Table 6.2 *FY95 Implementation plan: Quality service through quality training*

Prepared by: Karen-Prior Smith		Date 05/05/95	Rev: 1	Entity: UKSR				Department: Finance & Remarketing									Remarks ○ On track Warning ● Off track
No. Strategy	Step	Tactics	Owner	Q2	Q3			Q4			Q1 1996			Q2 1996			
				A	M	J	J	A	S	O	N	D	J	F	M	A	
1 To provide a professional, pleasant friendly service which maximizes our ability to conduct business over the phone through the effective use of our systems and maximizing our productivity.	1	Process binder mapped	CB		X												○
	2	Use of systems on-line capability identified	KS		X												○
	3	Company X training brochure to be customized to HP Finance	PW					X									
Performance metrics	4	Trainers identified	KS					X									
•Training plan in place •Trainers identified •Programmed devised	5	Review process determined	KS					X									
	6	Training conducted	TBA								X						
	7	Performances feed back	TBA								X						
	8	Ongoing training plan implemented	KS									X—	—	—	—	—	
Implementation plan		File name	Issue No. 1						Date 05/05/95						Page 1 of 1		

Table 6.3 *Quality service through effective complaint handling*

Prepared by: Karen-Prior Smith		Date 05/05/95	Rev: 1	Entity: UKSR			Department: Finance & Remarketing										Remarks ○ On track Warning ● Off track
No. Strategy	Step	Tactics	Owner	Q2	Q3			Q4			Q1 1996			Q2 1996			
				A	M	J	J	A	S	O	N	D	J	F	M	A	
2 To have a process for complaint handling	1	Define 'complaint'	All		X												○
	2	Document process and follow-up letter	CB/PW		X												○
	3	Monitor via Lotus Notes	CB/PW		X												○
Performance Metrics	4	Action plans in place to reduce volume	CB/PW			X											
100% follow-up on every complaint	5	Train group on new process	CB/PW			X											○
	6	Provide 'aide memories' and visible reminders of how to handle complaints	PW			X											○
Implementation plan		File name	Issue No. 1						Date 05/05/95								Page 1 of 1

Table 6.4 *FY95 Implementation plan: Quality service through quality recognition*

Prepared by: Karen-Prior Smith		Date 05/05/95	Rev: 1	Entity: UKSR	Department: Finance & Remarketing											Remarks ○ On track Warning ● Off track
No. Strategy	Step	Tactics	Owner	Q2	Q3			Q4			Q1 1996			Q2 1996		
				A	M	J	J	A	S	O	N	D	J	F	M	
3 To effectively reward and motivate staff who demonstrate service level beyond the call of duty	1	Manager to have 3 Golden Envelopes per month containing incentives	KS		X											○
Performance Metrics	2	Award 'on the spot'	KS		X											○
Feedback from employees	3	Advertise service contributors	KS		X											○
Implementation plan		File name	Issue No. 1						Date					Page 1 of 1		

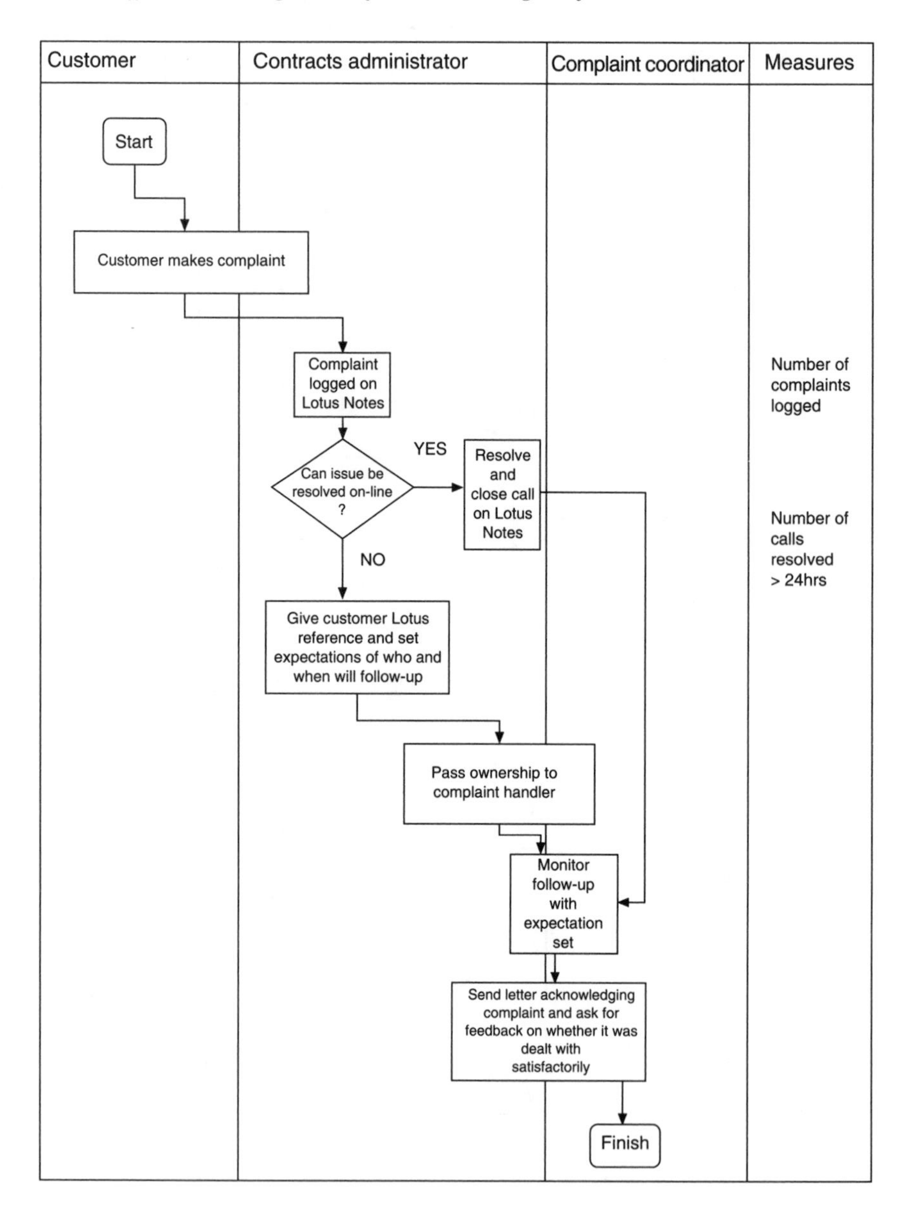

Figure 6.6 *The complaints process*

Figure 6.7 *Complaints handling traffic lights*

when you have customer satisfaction levels of 98–99 per cent there is still scope for improvement through benchmarking. Essentially this is because customers are asking the question of 'how do you know you are the best?' and at the IBM National Call Management Centre (NCMC) customer relationships management is a high-level core process. The NCMC is part of the solution delivery sub-process.

The NCMC is the first point of contact for IBM's customers requesting service on hardware, software and service products. The NCMC also have a General Enquiry desk available for both existing and potential customers. The NCMC was established in 1992 with a mission 'To provide "Best of Breed" Call Management Service ensuring value for money through our people, facilities and techniques'.

The NCMC has committed to answer 98 per cent of customers' calls to the NCMC within 10 seconds, with less than 2 per cent unable to complete the call. Customer satisfaction, as measured by an external and independent organization, has been 98 per cent for the past two years and increased to 99 per cent during 1995.

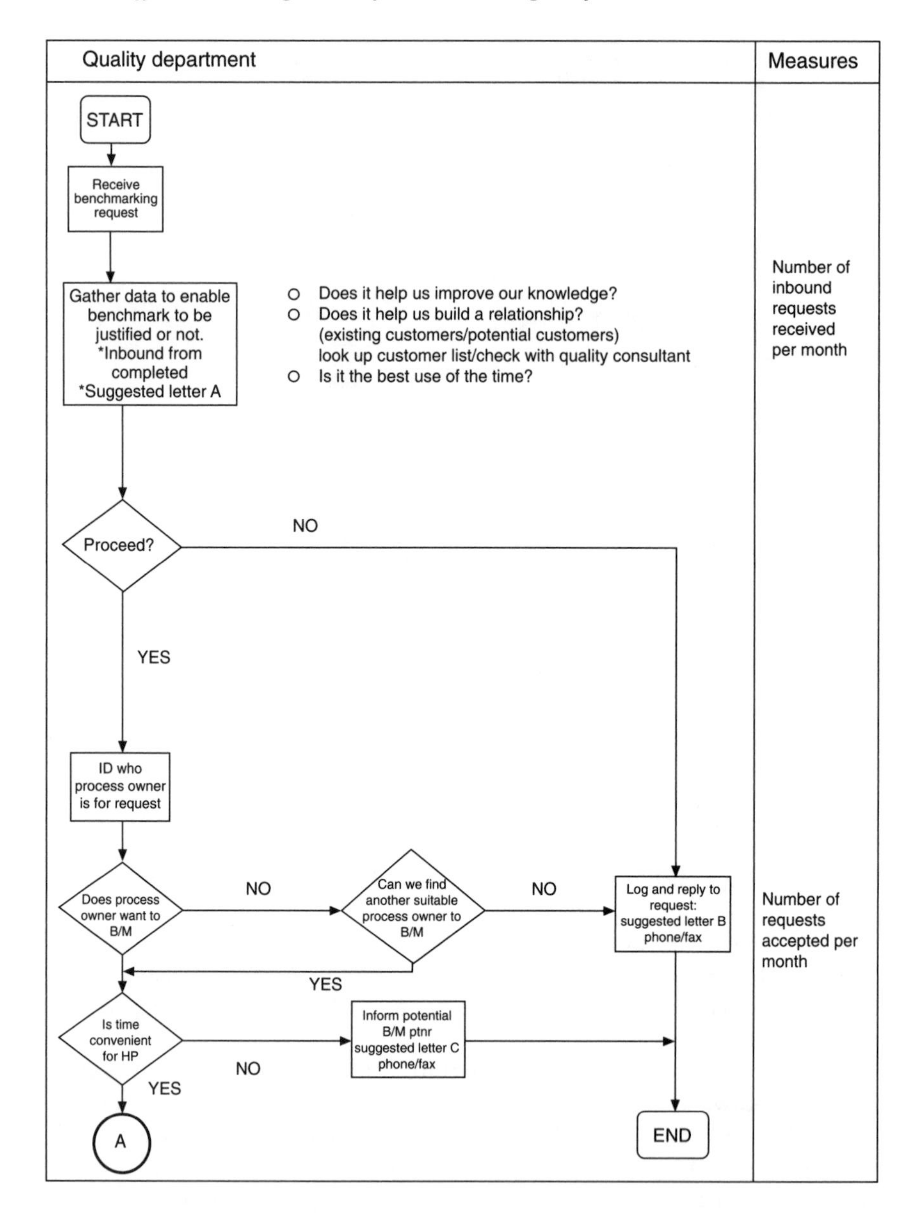

Figure 6.8 *Inbound benchmarking process with HP*

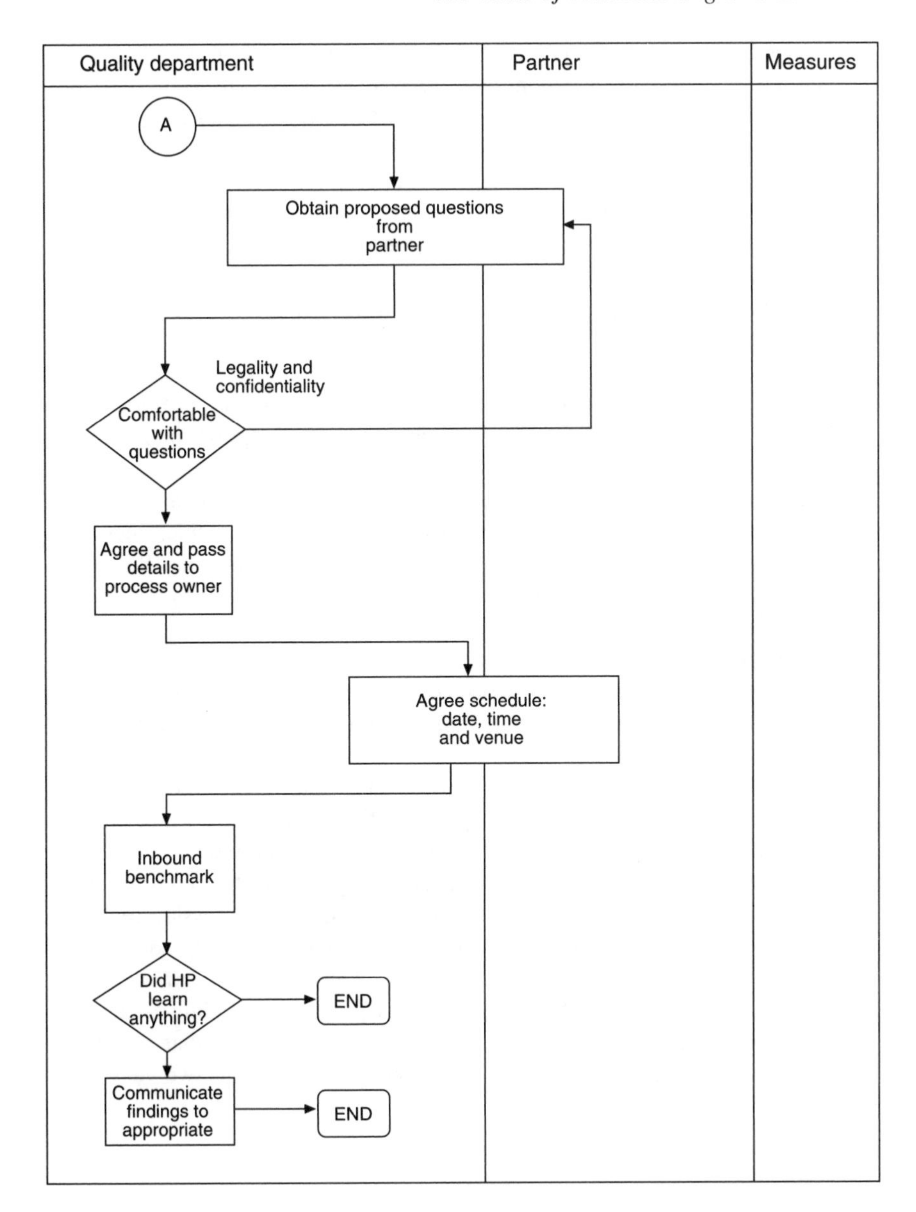

Figure 6.9 *The inbound benchmarking process within HP (continued)*

The NCMC has applied an innovative approach in its design and integration of human resource management and service delivery, ensuring that the 'moments of truth' – the primary interaction with IBM's existing and potential customers – is of the highest quality possible. A unique partnership has been created between IBM UK Ltd and Manpower plc, where the partnership created will ensure that where both stakeholders will benefit from any savings resulting from improved processes and increased value for money for our customers. Manpower plc, who are specialists in recruiting highly skilled customer service staff, complement IBM's skills in recruiting process and systems professionals.

The NCMC organization chart (Figures 6.10 and 6.11) reflects the Manpower/IBM relationship, where Service Delivery is provided by Manpower's resource and Supplier (Technical/Systems Support) and Business Development are provided by IBM. All these processes are mapped and underpinned by more detailed work-instructions. This is manifested in a high level of contracted staff, approximately 93 per cent, at the NCMC, which includes all Call Agents (telephone operators), and Service Managers. The IBM staff (eight Managers and Professionals) are seen as supporting roles to the operational (contracted) staff (see Figure 6.12).

The contract itself is seen as innovative, as it provides a win–win situation for both organizations, as well as the customers, and it has therefore been used as a benchmark for other contracts being drawn up within IBM's purchasing department.

The NCMC's mission is to provide world-class call management service. Starting from a very high level of customer satisfaction we are using benchmarking as a major tool in improving our practices on a continuous basis. Tele Centre Management is a fast-developing and changing industry and it is therefore vital not to 'rest on our laurels' if we are to ensure our mission remains true and achievable. We use the *kaizan* approach of small, continuous improvements in our technology, human resource management and processes through benchmarking in order to maintain our position of leading-edge telecentre – both request for best practices. It is on this basis that we have designed and applied our benchmarking activities over the past two years.

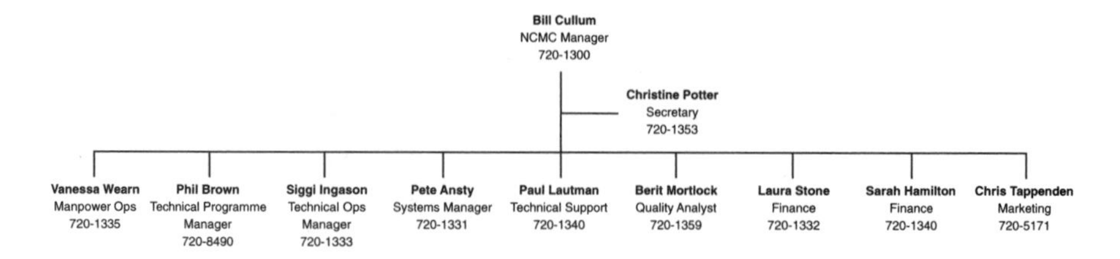

Figure 6.10 *The NCMC organizational chart*

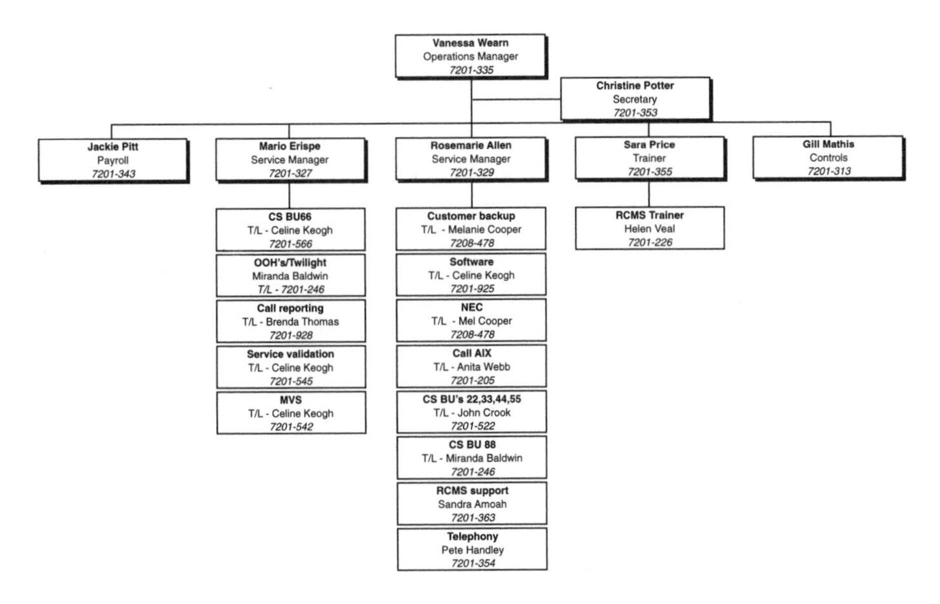

Figure 6.11 *The NCMC organizational chart – Manpower operations*

6.5.1 Why benchmarking?

The Baldrige self-assessment is promoted within IBM as the quality tool for identifying areas for improvements. In September 1993 the NCMC Management Team applied the Baldrige Model to measure the NCMC as a quality organization. It was clear that we were not able to justify our claim to be 'best of breed' – neither to IBM's customers nor to customers of the NCMC itself – until we had compared ourselves to other quality call management centres. And how about our business vitality? Telecommunications technology is developing fast, and how did we know that we could provide value for money to our customers, and remain competitive if the technology was overtaking us? The NCMC had been established in the spring of 1992 and we were now 18 months on. What improvements had taken place over that period both inside and outside the industry? What improvements had we effected within the NCMC? We did not have any positive answers. The outcome of the Baldrige assessment was the agreement to establish a benchmarking team within the NCMC.

6.5.2 The team

During October, November and December the benchmarking team was formed, underwent training, did research, looked for best practices, talked with other practitioners, explored databases on benchmarking, and investigated as fully as possible *how* benchmarking is undertaken to ensure we would

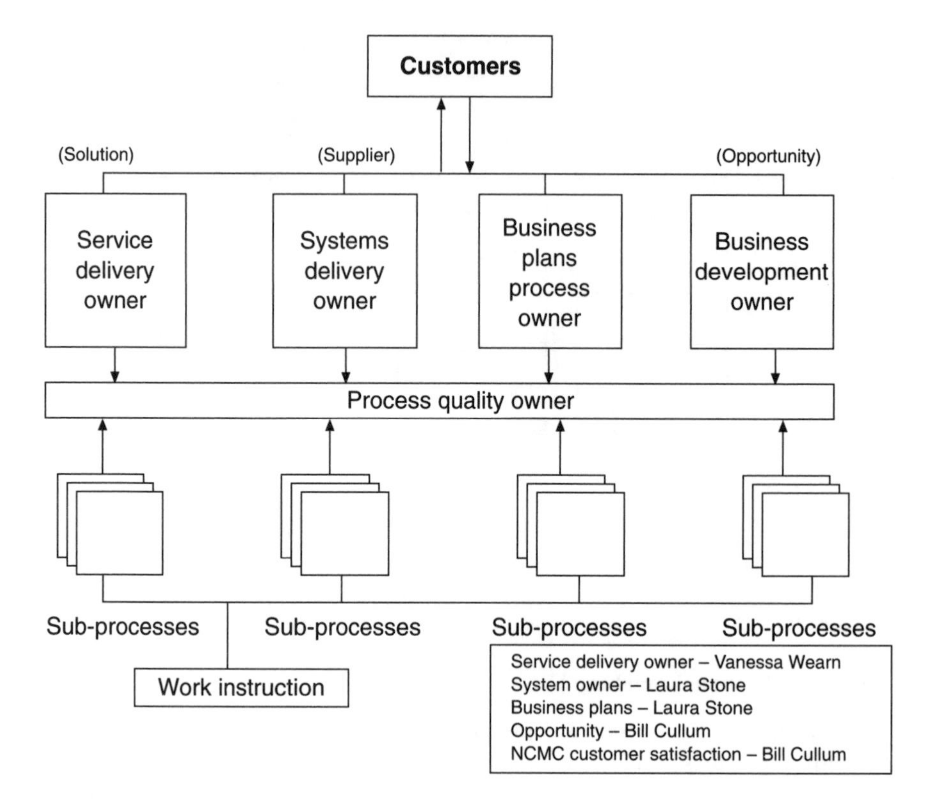

Figure 6.12 *The NCMC organizational chart – process flow*

gain from this activity. We were also aware that we would represent the NCMC – as quality ambassadors – to our potential partners. Our partners would also be giving us their valuable time, and it was therefore essential that we did our benchmarking training and research as thoroughly as possible before any visit took place.

To ensure as wide a covering of the operation and support provided by the NCMC the team members were drawn from:

Process Management – Berit Mortlock
Service Management – Vanessa Wearn
Operations Management – Heather Cross
Business Development – Jo Adams
Technical Support Group Management – Dave Hammett.

Only the team leader worked full-time within process management; the remaining team members were full-time practitioners within their field of work.

The selection of the team aimed to get as wide a representation of the NCMC process areas, also taking into consideration the skills and experiences of the team members. Responsibilities within the NCMC are as follows (see Figure 6.9):

- Seventy Call Agents
- Eight Technical Support Group Staff
- One Training Manager
- Three Operations Managers
- One Service Manager
- Four Technical/Telecommunications Managers/Professionals
- Three Process Management Professionals
- Two Business Development Managers
- Two Financial Management Professionals
- One Centre Manager.

It was considered important that the team was not too big, which would hinder efficiency, but at the same time it encompassed all the activities within the NCMC. It was felt that the five team members invited to the benchmarking team met all the requirements. The team consisted of a mix of managers and professionals, with full encouragement and commitment from the rest of the management team. The benchmarking team leader was also a trained facilitator, which ensured that the meetings were innovative and effective; as demonstrated in the approaches used as outlined below.

Diaries were cleared for 2 hours every Wednesday afternoon for the next six months for the benchmarking team to develop both as a team and to develop benchmarking activities that were appropriate to the NCMC. As a new activity it was felt that this resource was the optimum one both from the aspect of the size of the NCMC, the activities represented, as well as demonstrating its commitment to benchmarking.

6.5.3 Training

The first step for the team was to identify its training need, and contacted Barry Povey, the IBM Representative at the Benchmarking Centre, and requesting Benchmarking training from him. This was arranged, with Barry coming to the NCMC and providing a one-day benchmarking training workshop both for the team itself and the rest of NCMC management team.

The NCMC is based in Havant, where then IBM's disk-drive manufacturing plant was located. Our colleagues there had already been practising benchmarking both internally and externally, and we invited two members of their benchmarking team to our meetings for presentations and discussions on how they had undertaken their manufacturing benchmarking activities. These meetings were both valuable and interesting for us, as we were able to learn from them and their experience about their problems and lessons learnt before embarking on this journey ourselves.

IBM maintains an international databank on benchmarking activities, and through this we made contact with Gerald J. Balm, from IBM Rochester, Minnesota. We agreed to set up a teleconference with him, and through this media we had an inspiring and fruitful 2 hours, during which he shared his experience with us, provided practical advice, and filled us all with enthusiasm for benchmarking. He followed this up by providing each team member with a copy of his book: *Benchmarking: a Practitioner's Guide for Becoming and Staying Best of the Best.*

Following on from this teleconference Jerry Balm visited the UK a few months later and provided a one-day workshop at the IBM Education Centre for the IBM Benchmarking Steering Committee and members of the IBM UK Baldrige Implementation Team. Training for the NCMC benchmarking team also included reading of Bob Camp's book *Benchmarking: The Search for Industry Best Practices that lead to Superior Performance*. This was mandatory reading for each team member.

The IBM International Database on benchmarking activity was also searched for any benchmarking on telecentre processes. We found one which had been done by IBM in Dallas, with the Xerox Corporation in Rochester. We found this to be a good guide for a quality benchmarking activity, providing guidance on meetings, analysis, reporting and implementation of benefits.

By making full use of all the resources available to us, and still is, we were able to learn from the experience of our colleagues both in the UK and abroad; use databases and libraries for research both on how to start benchmarking as well as how it is put into practice. We also had access to an experienced trainer, and telecommunications facilities which we fully utilized.

6.5.4 Benchmarking practices

On the IBM benchmarking database we identified the benchmarking practices to which IBM adheres. It includes the Benchmarking Code of Conduct as adopted by the International Benchmarking Clearinghouse and the Strategic Planning Institute Council. All IBM benchmarking teams are expected to adhere to the following principles:

- **Principle of Legality**: Avoid discussions or actions that might lead to or imply an interest in restraint of trade, market or customer allocation schemes, price fixing, dealing arrangements, bid rigging, bribery or misappropriation. Do not discuss costs with competitors if costs are an element of pricing.
- **Principle of Exchange**: Be willing to provide the same level of information that you request, in any benchmarking exchange.

- **Principle of Confidentiality**: Treat benchmarking interchange as something confidential to the individuals and organizations involved. Information obtained must not be communicated outside the partnering organizations without prior consent of participating benchmarking partners. An organization's participation in a study should not be communicated externally without their permission.
- **Principle of Use**: Use information obtained through benchmarking partnering only for the purpose of improvement of operations within the partnering companies themselves. External use of communication of a benchmarking partner's name with their data or observed practices requires permission of that partner. Do not, as a consultant or client, extend one company's benchmarking study findings to another without the first company's permission.
- **Principle of First Party Contact**: Initiate contacts, wherever possible, through a benchmarking contact designated by the partner company. Obtain mutual agreement with the contact on any hand-off of communication or responsibility to other parties.
- **Principle of Third Party Contact**: Obtain an individual's permission before providing their name in response to a contact request.
- **Principle of Preparation**: Demonstrate commitment to the efficiency and effectiveness of the benchmarking process with adequate preparation at each process step, particularly at initial partnering contact.
- **Principle of Completion**: Share information about process, and consider sharing study results. Offer to set up a reciprocal visit, conclude meetings and visits on schedule.

As an IBM benchmarking team expecting to abide by this code we would request any benchmarking partner to agree to this code as well. We believe it encourages openness and trust within benchmarking activities.

6.5.5 Process

Before any benchmarking activity can commence it is essential to have defined and mapped the processes to be Benchmarked. IBM has defined process management (see Figure 6.12) as the ongoing methodology for assessing, analysing and improving the performance of key business processes based on customer needs and wants; with process management consisting of the following three elements:

- Process assessment involves determining customer requirements, mapping the process, collecting process performance data and then rating the process based on these data.
- Process analysis uses the data collected in process assessment to evaluate and develop plans to improve the process.

- Process improvement involves implementing the process improvement plan, obtaining customer feedback on the results, revising the plan as appropriate, and then implementing the solution.

They are part of a continuous process improvement programme, whereby a business's process performance can move from 'critical' to 'best-in-class'. The process model flow (Figure 6.14) demonstrates the importance of benchmarking – it is the first step within the 'process analysis' element. The NCMC benchmarking team leader had undergone process management training based on this model.

The benchmarking step itself has been defined in a process model flow and the benchmarking process contains four elements:

- Organization and planning
- Data collection
- Analysis
- Action.

How to perform each step within these elements is part of the benchmarking training provided within IBM, and which was provided for the NCMC benchmarking team and management.

The key process within the NCMC is call handling, i.e. from receipt of call to closing the call. The work instructions for this activity were well documented, and it was a clear candidate for benchmarking. Could we improve this one, as well as any of its subsidiary processes, e.g. service validation, first level system support? We were also keen to explore the human resource aspect of call management. Call receipt can become a monotonous activity with calls

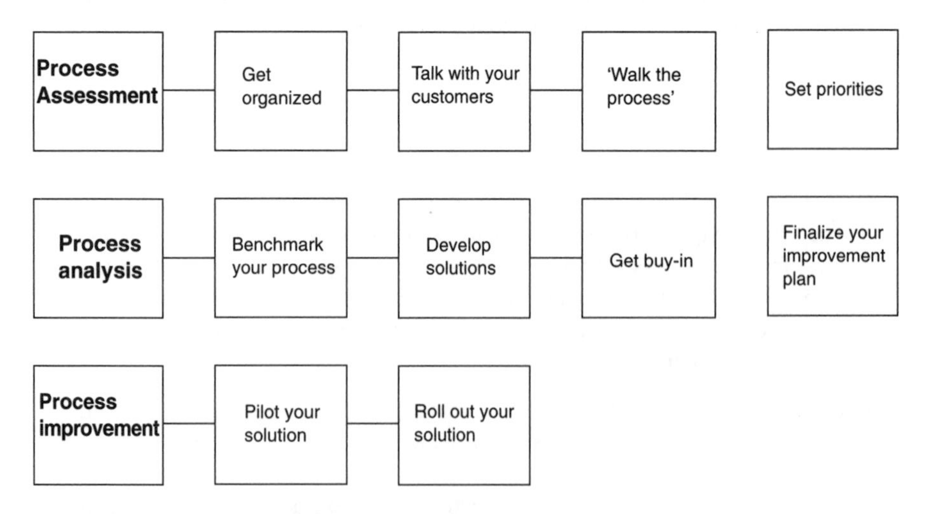

Figure 6.13 *IBM process improvement methodology*

being handled in quick succession by the call agents and motivation and customer care (our 'moments of truth') would suffer if this aspect was not managed well. It was agreed that these areas were key to our operation and to customer. By January 1994 the NCMC benchmarking team felt that the foundation had been laid for successful benchmarking activities by the NCMC and its potential benchmarking partners – both in the short and the long term.

6.5.6 Initial benchmarking activity

Selecting the benchmarking partner is an important step in benchmarking, as the 'wrong' – for whatever reason – partner can determine the success or failure of the activity. The NCMC team did not deliberate long in agreeing that Rank Xerox would be an ideal partner for us:

- As Bob Camp's book had demonstrated, it was a quality organization.
- Business vitality was demonstrated through its financial success.
- As Baldrige Award winners in the USA, they were familiar with this quality model and its requirements – aiming for quality goals within the same framework as IBM.
- They were a multinational company, providing service on their products – also similar to IBM.
- They understood the concept of and had experience of benchmarking, and would therefore understand and adhere to the Benchmarking Code of Conduct.

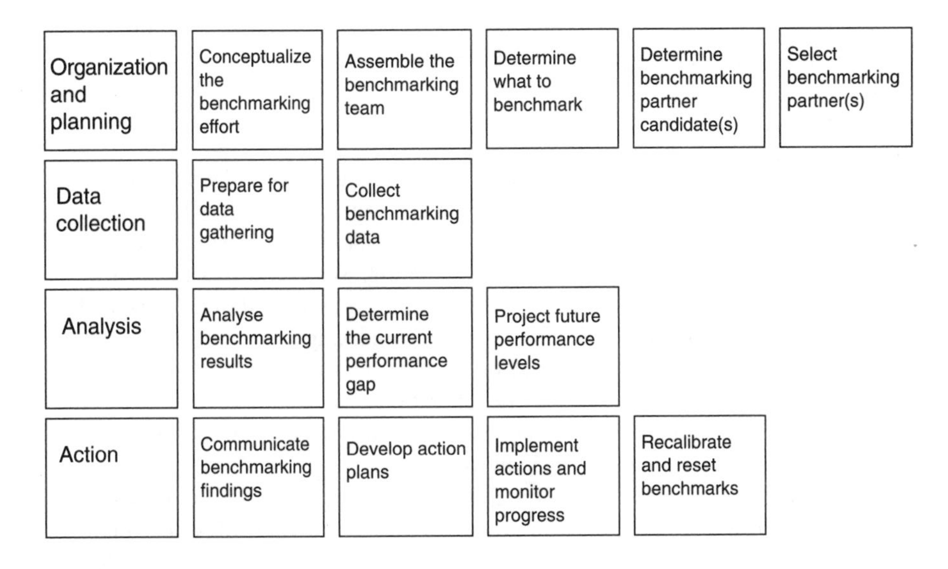

Figure 6.14 *IBM benchmarking process methodology*

- Familiarity with their telecentre had been established through Rank Xerox and the NCMC's involvement in establishing National Vocational Qualifications for telephone operators.

Rank Xerox welcomed us as their benchmarking partner, and the first visit was agreed to be at their Customer Response Centre in Milton Keynes and was arranged to take place on 1 February 1994.

6.5.7 Data collection

Rank Xerox was an excellent benchmarking partner. They understood the concept and had had previous experience, they were fully cooperative, open and sharing; answering all our questions. The NCMC team agreed on the questionnaire for each visit, and these were sent to Rank Xerox before the visits took place.

All team members were included in the site visits to Milton Keynes. The visit would commence with a plenary session where the general aspects of our questions would be answered, and we would then go off individually with Rank Xerox members of staff for more detailed questioning in our specialist areas.

On our return to Havant we would meet the following day and collate all information while it was still fresh. Over the next couple of weeks, we would analyse the data and information obtained, write the report and share this with Rank Xerox. During this period we also hosted two visits from Rank Xerox when we were able to reciprocate their hospitality. Through benchmarking we had established a partnership with fellow practitioners within an external organization with similar aims and values – i.e. business vitality and continuous improvement in a highly competitive environment.

6.5.8 Analysis

To avoid falling into the trap of 'industrial tourism' analysis of the information obtained from the benchmarking activity is key to its success. It would not be unfair to state that initially the analysis of the wealth of notes on the processes we had compared proved somewhat daunting. We found that the benchmarking experience we had learned from before going to Rank Xerox had been manufacturing based. We learnt that benchmarking processes in a service environment is different from that of a manufacturing environment. After comparing the statistical side of the operation – i.e. number of calls per week/answered within a number of seconds by a number of staff – the processes themselves were more difficult to evaluate.

However, we agreed to use Jerry Balm's book, and used the Spider Chart (Figure 6.15) in his book to illustrate our findings. The chart illustrates eight key measurements of interest, where the outer circle is the point where one has Total Customer Satisfaction (Ultimate Goal), and the centre point is 'barely any awareness' of the activity/process. The benchmarker will indicate where he has assessed his own current activity, as well as that of his partner. This graph will then illustrate where he is, with eight activities, compared to his benchmarking partner, on a scale of hardly any awareness to total customer satisfaction. For processes and activities within the service industry we have found this gap analysis and illustration invaluable. Within a manufacturing environment there are other quantitative analysis which are easier to apply, but less so in the service sectors.

6.5.9 Rank Xerox visits – benefits

Our benchmarking activity with Rank Xerox encompassed comparisons of

- **People** – Training/motivation/self-managed work group
- **Systems** – call receipt to close call – process overview
- **Customer satisfaction** – service level targets/measurements/process
- **Change management** – Communication.

In conjunction with Rank Xerox we agreed in our analysis of the findings that Rank Xerox exceeded the NCMC in the human resource management processes, particularly with regard to training and self-managed work groups. Within the other areas benchmarked, it was also agreed that the NCMC was on par or exceeded Rank Xerox. Our findings were shared with Rank Xerox and formally reported back to Rank Xerox in a document.

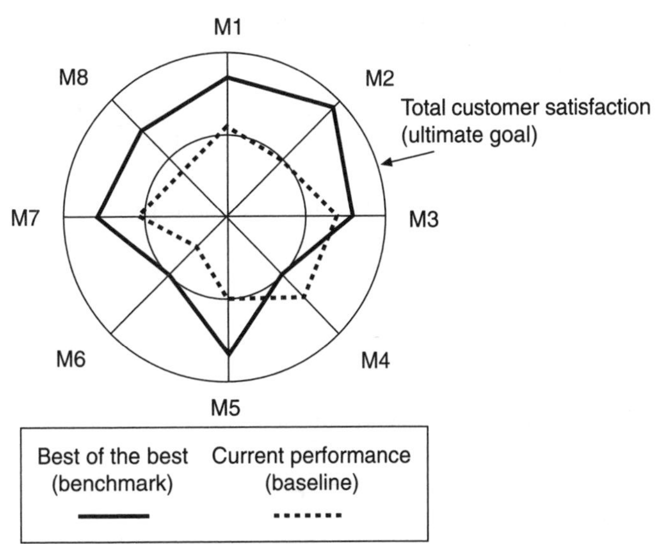

Figure 6.15 *Gap analysis on customer satisfaction – using the Spider Chart*

The NCMC has evaluated internal and external training and ongoing development of the specific need the Centre has in implementing self-managed work groups. This is a long-term culture change which includes significant role changes for team leaders and service managers. We consider it vital to our future success that we get this aspect of our operations right, and the training, coaching and ongoing support must be sensitive to our requirements and flexible in its approach and long-term support. This is now underway, and we expect to see improvement in employee satisfaction at our next survey, 14–28 June 1995.

The customer calling IBM at the NCMC identifies his response here with that of IBM. It is therefore crucial that the image presented and the service given is of the highest quality. We are therefore investing significantly in training and employee motivation for the call agents.

Training for all agents, (e.g. induction, customer care, telephone handling, systems, refresher, problem solving, etc.) has been reviewed with external training organizations presenting what they can offer in call management training. After a thorough review of four external companies, we have now engaged one of these to provide what we can consider after this evaluation the 'best of breed' call management training for our call agents.

The measurement of these benefits we consider to be in the future: by maintaining our position as a leading-edge telecentre by high morale, committed and highly trained staff, which is essential to maintaining our high customer satisfaction rating.

6.5.10 Benchmarking maturity

Process management

In July 1994 the NCMC embarked on the redesign of the IBM/Manpower contract for the NCMC operation. Roles and responsibilities had to be clearly defined, and an innovative win–win contract was formulated and agreed. From this work the NCMC was identified as a process-led organization (Figure 6.11). The four overall processes are:

- Service delivery
- Systems delivery
- Business plan
- Business development

with customer satisfaction the sum of all these processes, and therefore the responsibility of the NCMC manager.

The service delivery's key process, manage request, consists of five subprocesses:

- Receive request
- Perform entitlement check
- Secure response owner

- Monitor the request
- Report activity.

These are again supported by work instructions.

The business re-engineering which the NCMC underwent during the autumn of 1994 moved it from a hierarchy organization to a flat, process-driven one. It was influenced by innovation and benchmarking, e.g.:

- **International** – IBM Corporate initiate drive for global, streamlined processes.
- **National** – IBM UK Executive Board sponsoring the UK Benchmarking Committee; drives the company to achieve IBM's Baldrige Bronze Award (and Silver for 1995); require all key company processes to be processed managed according to the IBM process management model, which includes benchmarking.
- **Local** – the NCMC's redesign of its manpower contract – encompassing human resource management including training; together with benchmarking activity, especially with Rank Xerox where the human resource management processes were found to exceed the NCMC's.

Change in benchmarking approach

The role of the original NCMC benchmarking team changed as our benchmarking culture matured. It was agreed to end the weekly, minuted meetings in the summer of 1994; however, to continue to work with other benchmarking partners who requested visits to us, to provide advice and guidance to other areas within IBM, and to continue to develop our partnership with Rank Xerox. The NCMC also agreed to continue to be represented on the IBM Benchmarking Steering Committee where its experience could be shared and other company benchmarking activities supported.

Individual team members provide benchmarking training and support to any member staff of the NCMC who are involved in process improvement activity, and has specialist knowledge in their field. Original team members provide ongoing training and support and act as a centre of competence for ongoing benchmarking activities.

This change reflects the process awareness; the knowledge that it is everybody's responsibility to look for best practice; and that the practitioner is the ideal person to compare and improve. We now have a centre of competence in place, to be drawn upon as and when required.

Benchmarking form

With the maturing of our benchmarking activity and benchmarking processes the NCMC has developed a benchmarking form which covers all aspects of our business, including:

- Mission/vision
- Call centre statistics
- Organizational infrastructure
- Technical infrastructure – telephony
- Technical infrastructure – systems
- Support
- Security
- Customer satisfaction
- People/human resource management
- Environment
- Manage request process.

These forms are designed to cover every activity within the NCMC, except financial. The latter is usually outside the scope of any benchmarking activity. It is designed to enable us to understand how close to 'apple for apple' comparison we are undertaking. The forms are also flexible enough to enable a tailoring to a specific organization, or just one or two areas can be pulled out for specific visits as it is rare that a complete telecentre benchmarking takes place. Individual processes are more usual. A copy of one page of the benchmarking form is shown in Figure 6.16.

We are using these forms to benchmark the NCMC with itself – a complete set was completed in February 1995 on where we are today, and what our processes are. The same activity will be undertaken in February 1996 and we will compare the two. We expect to see continuous improvements in all areas to ensure process vitality.

We are also using the NCMC staff's business calls – made from their workstations – to evaluate the criteria of the response they receive against the cri-

IBM:			Page 2 of 2
Organization:			
Benchmarking topic: Call centre statistics			
Interviewer		Interviewee	File name
Date		Position	
Sub-topic	**Description**	**Approach**	**Deployed**
	• Identify type of call the above refers to, e.g. inbound, enquiry direct, etc.	....	
	• Call distribution (by hour/day/month)	....	
	• Number of calls agents	....	
	• Number of turretts	....	

Figure 6.16 *Benchmarking collection proforma used by IBM staff*

teria by which we evaluate our own service delivery. Forms were designed for the staff to use and to be completed during and on completion of the call. This benchmarking activity commenced in May 1995, is continuing, and the initial results are currently being analysed.

This evolution and maturity of our benchmarking activity and process management caused us to reassess and re-evaluate our benchmarking activities further. We knew it was important to maintain our process vitality to ensure leading-edge business practices and to maintain high customer satisfaction. We felt comfortable and could demonstrate our expertise, and we therefore decided to channel this into three specific areas:

- The TeleBusiness Association
- Internal IBM comparisons – both national and international
- Sharing with other innovative UK businesses.

6.5.11 The Telebusiness Association

The NCMC was invited to join the TeleBusiness Association (TBA) Benchmarking Forum as a founder member, on behalf of IBM. We are now part of the TBA Steering Committee as well as Benchmarking Group 2. The main objective of the TBA is to act as a platform for structured networking and benchmarking between its members.

The TBA's Benchmarking Forum aims to achieve a set of common standards in telebusiness, providing support for organizations interested in identifying and implementing best practice. The Forum is split into working groups, each of up to ten non-competitive organizations. Participants are selected on the basis for their contribution and influence. The founder members, which include Coca-Cola & Schweppes, Forte Hotels, National Westminster Bank Next Directory, Private Patients Plan, Virgin Atlantic and the NCMC, have been working to establish key areas of best practice for telebusiness since March 1994.

In an emerging industry like telebusiness, one of the greatest challenges is to arrive at widely recognized and easily adopted set of standards for best practice. These must be geared towards business needs and updated on a continuous basis to keep up with advances in technology and ideas. We see our support of the TBA and its Benchmarking Forum as key to our meeting this challenge.

6.5.12 Sharing

Sharing is synonymous with the culture fostered by benchmarking – sharing our best practices with other businesses and organizations in the UK and abroad. The NCMC firmly believes in the benefit of benchmarking and sharing experiences with other businesses and are hosting visits to external companies on a regular basis – an average of two or three per week. Over the last two months we have hosted the following:

- GEC Marconi
- Barclays Merchant Services
- Ladbroke Racing Ltd
- Coca-Cola & Schweppes
- PPP Personal Insurance
- RAF
- Federal Express
- Coop Bank
- Mercury
- General Accident
- Sun Alliance
- Next Directory
- CSA
- Techmark
- The Prudential
- Royal Life
- Zurich Insurance
- Norwich Union
- Security Express
- Portsmouth Water Company
- Motorola
- Abbey National
- Friends Provident
- Eagle Direct
- Forte
- British Telecom
- British Rail
- Oslo Power, Norway
- IBM Italy
- IBM Japan

6.5.13 Internal benchmarking activity

As IBM UK and worldwide is progressing in its achievement of the IBM Internal Baldrige Award, we are finding valuable benchmarking partners within our own organization. IBM Malaysia was one of the first IBM Units outside the USA to achieve the highest Baldrige Award. Having read their Award Submission it was clear that they had achieved impressive results, e.g. overall company customer satisfaction improvements from 54 per cent to 94 per cent over the last 4 years. One of the factors quoted as being responsible for this improvement was the establishment of their Customer Relations Centre (CRC) with its Single Point of Entry (SPOE). This included the telephone contact which customers made when requiring service for their hardware and software products, i.e. the service provided by the NCMC. This indicated that they had a process in place for customer calls which the NCMC might learn something from.

The NCMC Quality Process Owner contacted IBM Malaysia in January 1995 and requested to benchmark their telecentre process. They were happy to agree and a visit was made to their location in Kuala Lumpur in February this year. The benchmarking forms designed at the NCMC were used as a basis for the comparison.

The Malaysian operation is on a much smaller scale than the UK one. However, the concept of customer care is pervasive within their operation and a pleasure to experience. The Technical Service Desk was on par with the UK in providing rapid response, and lost call rate. However, their technological platform was less advanced than ours, and the owner of this operation in Malaysia has requested a visit to the NCMC to gain understanding of our advanced expertise, processes and systems in this area. Truly a win–win activity.

We obtained information about their excellent customer care and SPOE activities, as well as customer satisfaction programmes. This information we have passed to the relevant part of IBM UK, and will as a result be actively participating in a project with a view to a similar customer care programme being established within IBM UK. This project commences 15 June 1995.

NCMC call management systems specialists and business development specialists visited IBM Sweden, France and Italy during the first half of 1995. All were given benchmarking training before the visits and walked through the IBM Benchmarking Process and Code of Conduct. Contact was made with the partners in the European units; questionnaires were mailed prior to the visits, the NCMC benchmarking forms were used; and reports were filed on the return to the UK. Analyses were made from the findings; and recommendations were made, e.g.:

- Continue with emphasis on measure and improve the longest wait time for calls, rather than driving down an already keen response time
- Review using field businesses for back-up rather than separate back-up facilities
- Separate cost of systems from the other running costs of the NCMC
- Helpdesk availability for customers without maintenance contracts via separate telephone number.

These are some of the findings from the trip to IBM Sweden in June 1995. The recommendations are not revolutionary, but will ensure process vitality, continuous improvement and an innovative approach to all our processes. It will enhance our internal partnerships to ensure our customers – national and international ones – are satisfied with our service delivery.

6.5.14 Key activities

The NCMC is able to offer our internal and external benchmarking partners the benefits of leading-edge technology and resource modelling within a telecentre. In April 1995 we commenced implementation of Computer Telephony Integration; and we plan to complete this by the end of September 1995. This technology provides the call agents with all the customer information as the telephone call is answered, without having to request this from the caller. It also enables the agent to pass the call to a specialist together with the customer information on the screen – thus eliminating the need for the caller to repeat all his or her details to the next person. This is an important step in maintaining customer satisfaction.

Resource modelling is also fully computerized, using the latest hardware and software. It helps the NCMC plan the resource according to the call demands, i.e. meet the peaks and manage any excess resource efficiently and effectively. It will also ensure that sudden illness is managed and the customers will not experience a decrease in the service we provide.

We believe benchmarking is an activity to promote best practice and to ensure mutual benefits. The NCMC is keen to ensure that our benchmarking partners will return from their visits to us knowing it was of value to them.

6.5.15 Results

The NCMC has introduced activity-based costing. Being process managed, we can evaluate the cost of each activity within the NCMC. The focus is therefore on the cost of each process, and the opportunity to improve the financial aspects of processes by improving the processes is an excellent incentive for benchmarking.

The vitality of the business is equally as important as customer satisfaction. If this balance of efficiency and effectiveness is out, the business will suffer in the long run. We are therefore very keen to maintain our customer satisfaction results of 99 per cent, and see benchmarking as the key to maintaining this balance.

6.6 The value of the European Best Practice Benchmarking Award

In addition to the obvious benefits of receiving an accolade and a certificate, this benchmarking award is perhaps the best way of recognizing the efforts of project teams and of promoting best practice and even entrants who do not wish, get feedback information from the judging panel.

As far as HP and IBM are concerned, several unexpected benefits did materialize:

- The external recognition has motivated the teams concerned to go on exploring and improving in other areas.
- The expertise gained through implementing the benchmarking process has helped corporate organizations for the transfer of skills and the facilitation of other projects undertaken.
- Externally, it helped create a large network of contracts and access to various sources of information.
- It helped promote the corporate brand names as two excellent organizations where best practice is abundant.

As the team at NCMC summarized their experience: 'This constant focus on benchmarking and best practice has, we believe, left the NCMC with a benchmarking best practice culture of which we are very proud.'

Acknowledgements

The case studies from Hewlett-Packard and IBM were reproduced with kind permission from MCB University Press Ltd.

7 The process of benchmarking in practice

Wisdom denotes the pursuit of the best ends by the best means

Francis Hutcheson

The task of the leader is to get his people from where they are to where they have not been

Henry Kissinger

It is not goodness to be better than the worst

Seneca

7.1 Building a culture of benchmarking

Companies that are very serious about the institutionalization and embedding of benchmarking take serious steps to ensure that what is being focused on is not necessarily just the project in question and its desired outcomes. Indeed, in many cases there are substantial resources being allocated to monitor the deployment of corporate benchmarking, the learning resources available, the initiatives and efforts for creating awareness through training, the incoming requests for participation in initiatives and the outgoing efforts and their progress.

Rank Xerox Ltd, for instance, monitor the level of pervasiveness of benchmarking through their self-assessment process. Using a framework referred to as the 'Business Excellence Certification' Model, individual business units are rated yearly on a scale of 1–7 on how pervasive their benchmarking efforts are. As Figure 7.1 illustrates, there is clear progression in most of the European business units concerned.

The Rank Xerox Ltd approach is indeed dynamic since benchmarking drives policy and strategy and targets for the 'vital few' areas are only determined through gaps in performance identified from benchmarking activity both internal and external.

To the question: 'How far has Rank Xerox managed to get Benchmarking embedded into its culture? Mr Paul Leonard, Senior Consultant at Rank Xerox Quality Services Ltd gave the following answer:

> . . . on a scale of one to seven where *one* is no application of benchmarking, no understanding of benchmarking, through to seven which is world-class performance, where benchmarking plays an integral part in the overall management

	1	2	3	4	5	6	7
1991		1	9	11			
1992			8	14			
1993			1	5	13	2	

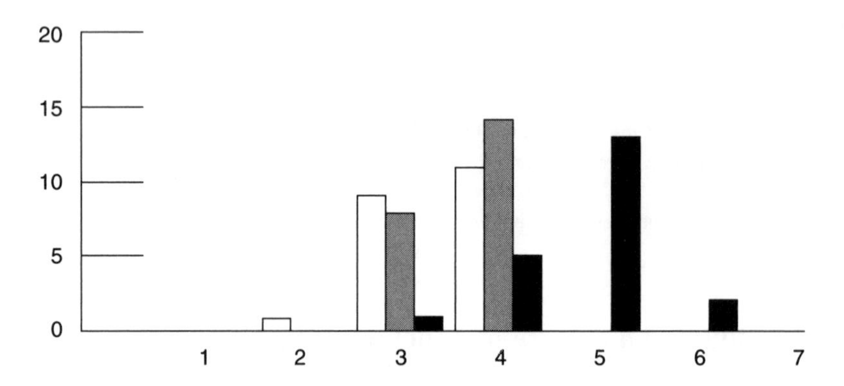

Figure 7.1 *Benchmarking pervasiveness at Rank Xerox*

> process including the strategic planning of the organization, then I would guess generally across the whole of Rank Xerox we would be between five and six. Having said that there is still a lot of work to be done to get all the units up to world-class standard and using benchmarking as it is currently known.

Most organizations in the West have greatly been inspired and influenced by the Xerox methodology. Some have adapted it extremely well to suit their own individual cultures and to ensure that the practice of benchmarking can yield to consistent positive outcomes and will enhance their competitiveness. It is therefore very worthwhile examining some of the best in class processes used by model organizations.

7.2 IBM

Benchmarking is presented within IBM business units worldwide as 'an ongoing discovery and learning activity for improving key business processes'. It is defined as follows:

> The continuous process of analysing the best practices in the world for the purpose of establishing and validating process goals and objectives leading to world-class levels of achievement.

IBM has developed its benchmarking process with the following objectives in mind:

- To ensure that our business process goals are set to exceed the best quantitative results achieved by world-class leaders
- To incorporate best practices throughout the IBM business process
- To reach a level of maturity where benchmarking is an ongoing part of the management system in all areas of the business.

In order to support global corporate benchmarking activity, IBM introduced a benchmarking support strategy with the responsibility given to the senior vice president of market driven quality, IBM's approach to Business Excellence. The benchmarking support strategy is meant to provide consistency and a flexible approach worldwide and an effective information flow among sister companies. Its key objectives therefore are:

- To promote the maximum return on investment for all of IBM's benchmarking activities
- To provide a mechanism for enterprise processes, lines of business (LOBs) and geographies to coordinate their benchmarking activities
- To reduce/prevent redundancy resulting from multiple requests to a benchmarking partner for the same information or different internal units both planning external benchmarking for a common process
- To provide a focal point for benchmarking education
- To provide a central repository for benchmarking information and results;
- To ensure communication and reporting of benchmarking requirements and activities
- To ensure that IBM uses leading-edge benchmarking tools and techniques.

7.2.1 The benchmarking process

The IBM benchmarking process is based on proven practices used within Rank Xerox and AT&T. The model, which is illustrated in Figure 7.2 is represented by four major phases as fourteen steps;

Phase I Organization and planning
Phase II Data collection
Phase III Analysis
Phase IV Action

In order to ensure effectiveness of using the methodology and achieving successful outcomes, IBM developed a checklist for each of the four phases to be closely followed by the teams involved in benchmarking activity (Tables 7.1–7.4).

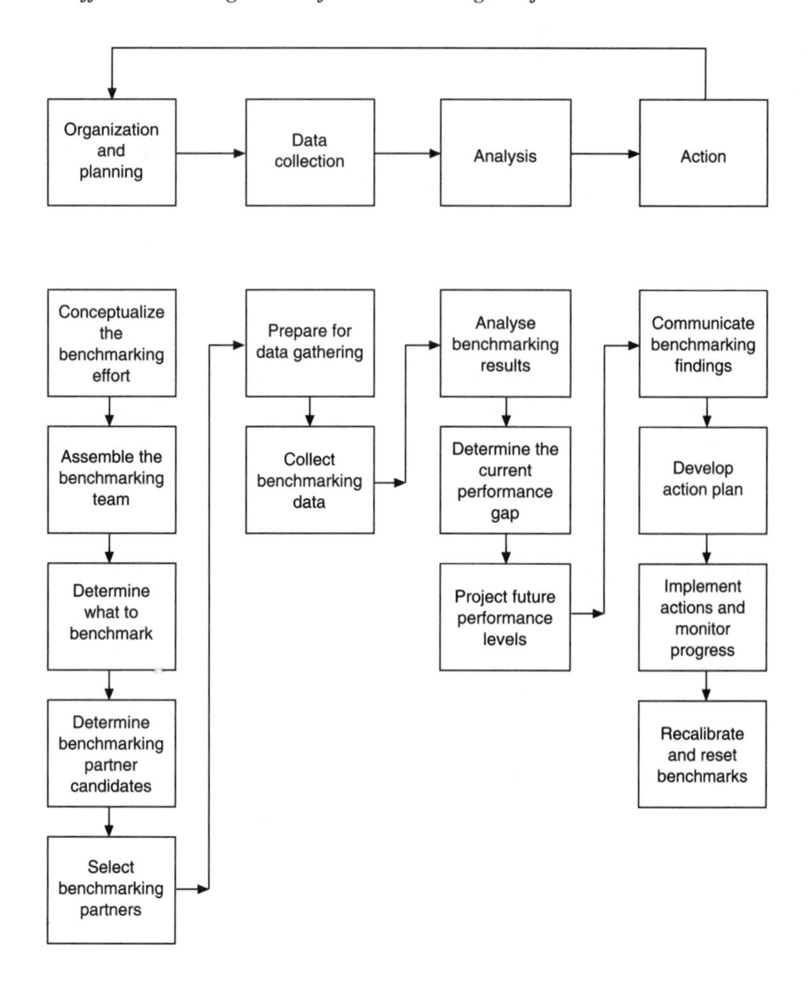

Figure 7.2 *IBM's benchmarking process model flow*

To ensure that a culture of benchmarking will ensue, IBM has devoted vast resources and tools, including, for instance:

- Benchmarking education
- Benchmarking reference materials
- Benchmarking electronic forum
- IBM Worldwide Benchmarking Registry
- Field Directory (customer accounts and geographical areas).

IBM refers to a variety of sources for benchmarking research and information acquisition. Among the chief sources and databases available are:

Table 7.1 *Phase I checklist*

Item	*Tick completion*
A complete analysis of the IBM process, produce or service to be benchmarked has been completed	☐
An overall benchmarking plan has been established	☐
The required resources, commitments and management sponsorship needed to successfully complete the benchmarking effort have been obtained	☐
A team that includes individuals with appropriate skill and knowledge of the business area being benchmarked has been assembled to plan and execute benchmarking activities	☐
At least one member of the benchmarking team has been fully trained in the benchmarking process	☐
Sufficient benchmarking education and skills have been acquired by team members to ensure effective benchmarking efforts	☐
The IBM Worldwide Benchmarking Registry (BMKC) has been checked to determine if any other IBM groups have been benchmarked or are planning to benchmark the same or similar processes	☐
The appropriate IBM benchmarking coordinator has been informed of the benchmarking plans	☐
An appropriate set of measurements, particularly customer-focused measurements, and specific topics have been selected for benchmarking	☐
Detailed research and investigation have been conducted to find world-class leaders in the business area being benchmarked	☐
The marketing team has been contacted for all benchmarking with current and potential IBM customers	☐
IBM legal counsel has been contacted about the proposed benchmarking activity and potential benchmarking partners	☐
The required procurement (purchasing) function has been contacted for benchmarking with IBM suppliers	☐
The IBM Worldwide Benchmarking Registry (BMKC) has been searched for other activity with potential benchmarking partners	☐
A good mix of benchmarking partners and alternates has been selected to provide a sufficient richness of benchmarking information	☐
The planned benchmarking activity has been registered in the IBM Worldwide Benchmarking Registry (BMKC)	☐

Table 7.2 Phase II checklist

Item	*Tick completion*
The data collection method and criteria have been selected	☐
IBM management and legal counsel have been informed about information to be shared with an external benchmarking partner, and any potentially confidential or sensitive information to be exchanged has been approved by them	☐
A clear, concise set of questions has been developed on the benchmarking topics	☐
All required site visit agendas have been created and discussed with benchmarking partners	☐
Information on practices and methods has been collected	☐
Information on measurements and their background details has been collected	☐
The correct amount and level of detail has been collected from all benchmarking partners	☐

- IBM Technical Information Retrieval Centre (ITIRC)
- Vendor electronic databases
- Professional associations
- Business libraries
- Outside consultants
- Questionnaires.

To round it all off, IBM has developed a series of questions to help benchmarking teams assess how well they are doing with the approach to benchmarking, and the results achieved. The list of questions includes:

- Do you use a systematic process for selecting comparisons?
- Is there a strong relationship between processes to be compared and MDQ goals?
- How thorough is your research to identify benchmarking partners?
- Do you compare against competitors and non-competitors?
- Can you demonstrate that the process comparisons and data sharing you perform are like those of best-of-breed and world-class operations?
- How many different criteria are used for comparisons and how appropriate are they?
- What is the scope (e.g. number of sub-processes) of your process benchmarking?

- Do you use outside sources of data and how often do you use them?
- Can you provide a clear explanation of how benchmarking data are used to encourage innovation and better knowledge of processes?
- Do you use benchmarking data to set goals and standards?
- Do you use benchmarking data to stimulate the identification of opportunities for quality improvement?
- How does your process output, methods and supplier quality compare with other processes considered best-of-breed or world-class?
- Have you achieved demonstrable improvements due to benchmarking?
- Have you achieved a process leadership position?
- Do you have a strategy for expanding the scope of future data collection?
- Has benchmarking become integrated into the daily activities of every portion of your business?
- Does everyone in the organization believe in, and are they committed to, benchmarking?

Table 7.3 *Phase III checklist*

Item	*Tick completion*
A debriefing of the benchmarking team was held as soon as possible following the benchmarking meeting	☐
A detailed analysis of the data has been completed	☐
Gap analysis has been completed for measurements collected	☐
Future performance trends have been determined	☐
The detailed analysis of the data has been completed	☐

Table 7.4 *Phase IV checklist*

Item	*Tick completion*
The benchmarking findings have been documented in a written report	☐
The IBM Worldwide Benchmarking Registry (BMKC) has been updated with the completion date, an abstract of the findings and a contact for obtaining the full report	☐
The results have been communicated to the organizations involved	☐
Action plans have been created and approved	☐
Actions are being implemented and monitored	☐
Benchmarking has become an integral part of the planning and business improvement process	☐

7.3 American Express

American Express recognizes the importance of benchmarking like many other model organizations. Amex often quotes a survey of the *Fortune 500* companies which has demonstrated that first, benchmarking has significantly grown over the past few years and second, that in twenty-four industries, companies that benchmarked outperformed non-benchmarking competitors by more than 7 percentage points in return on equity. Further, the importance of benchmarking is recognized by the fact that using the Malcolm Baldrige National Quality Award framework, benchmarking, along with competitor analysis, impacts more than 55 per cent of the total award criteria.

Amex recognizes the areas most important for benchmarking purposes:

- Business processes which provide a major source of competitive advantage to American Express
- Areas which strongly affect employee or customer satisfaction
- Areas with rapidly changing technology
- Practices where costs are felt to be non-competitive

7.3.1 The American Express benchmarking process

Amex has recognized the need to have a systematic methodology to benchmark. Based on the Rank Xerox approach, Amex has developed its own methodology, using five phases and fifteen steps (Figure 7.3).

To create a culture conducive to effective benchmarking, the following arguments are presented to remind people on taking the subject of strict, systematic benchmarking methodology seriously:

- Without a methodology, it would cost more in terms of time and money than the value of the resulting improvements.
- It is possible to waste resources by studying processes or functions that have already been benchmarked in a company.
- World-class organizations holding best practices may reject requests because of poor approach and preparation.
- Credibility problems and inability to persuade top management to endorse recommended change.

Through the use of a benchmarking council, benchmarking activities are coordinated among sister business division (internally) and with partner organizations (externally). Further, the role of benchmarking coordinators is to help improve the application of benchmarking and to develop appropriate expertise so that the standard of application is compatible with Amex guidelines and the accepted generic standard externally.

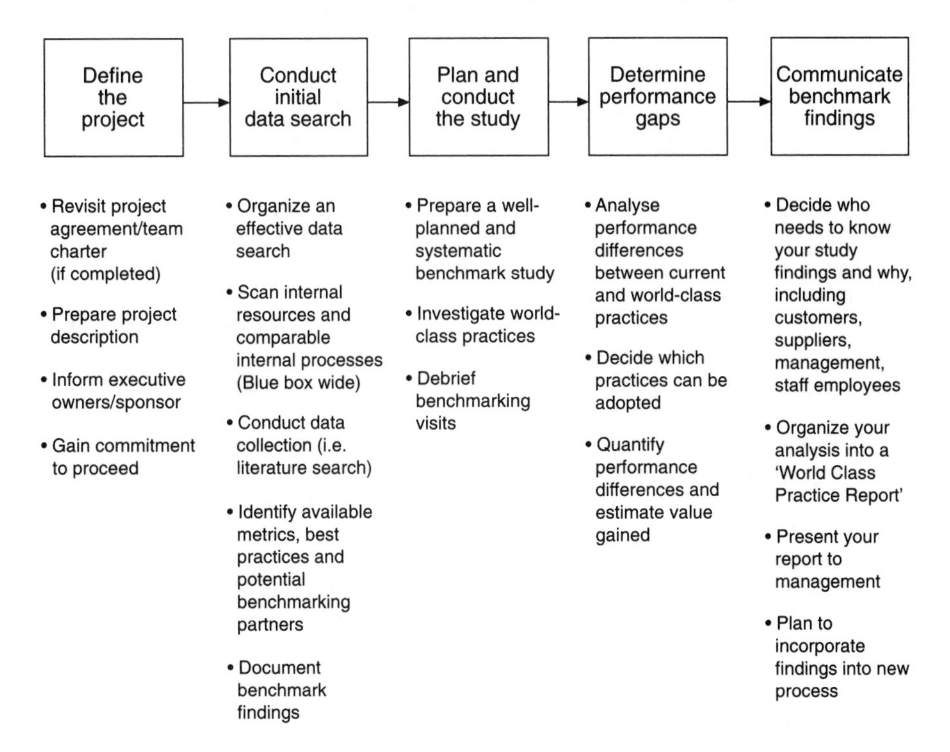

Figure 7.3 *The benchmarking process at American Express*

7.3.2 Managing inbound requests

Amex produced specific guidelines and an approach for handling external requests for taking part in a benchmarking activity. The following areas need to be thoroughly considered prior to a decision being made:

- Need to receive written description of the places/area of interest for benchmarking
- List of questions to be asked
- The list of people to be involved in the study
- Whether a site visit is expected or whether a straightforward telephone conversation can suffice
- What can be expected in exchange for Amex cooperating in the requested studies (for instance, a copy of a summary report/findings from the study; information sharing from visits conducted to other companies; the extended invitation to replicate site visit)?

Table 7.5

Etiquette and ethics

In actions between benchmarking partners, the emphasis is on openness and trust. The following guidelines apply to both partners in a benchmarking encounter:

- In benchmarking with competitors, establish specific ground rules upfront, e.g. 'We don't want to talk about those things that will give either of us a competitive advantage, rather, we want to see where we both can mutually improve or gain benefits.'
- Do not ask competitors for sensitive data or cause the benchmarking partner to feel that sensitive data must be provided to keep the process going.
- Use an ethical third party to assemble and blind competitive data, with inputs from legal counsel, for direct competitor comparisons.
- Consult with legal counsel if any information-gathering procedure is in doubt, e.g. before contacting a direct competitor.
- Any information obtained from a benchmarking partner should be treated as internal and privileged.
- Do not:
  - Disparage a competitor's business or operation to a third party
  - Attempt to limit competition or gain business through the benchmarking relationship
  - Misrepresent oneself as working for another employer.

Benchmarking exchange protocol

As the benchmarking process proceeds to the exchange of information, benchmarkers are expected to:

- Know and abide by the Benchmarking Code of Conduct
- Have a basic knowledge of benchmarking and follow a benchmarking process.
- Determine what to benchmark, identify key performance variables, recognize superior performing companies, and complete a rigorous self-assessment.
- Have a questionnaire and interview guide developed, and share these in advance if requested.
- Have the authority to share information.
- Work through a specified host and mutually agree on scheduling and meeting arrangements.
- Follow these guidelines in face-to-face site visits:
  - Provide meeting agenda in advance
  - Be professional, honest, courteous and prompt
  - Introduce all attendees and explain why they are present
  - Adhere to the agenda: maintain focus on benchmarking issues
  - Use language that is universal, not one's own jargon
  - Do not share proprietary information without prior approval from the proper authority, of both parties
  - Share information about your process, if asked, and consider sharing study results
  - Offer to set up a reciprocal visit
  - Conclude meetings and visits on schedule
  - Thank the benchmarking partner for the time and for the sharing

7.3.3 Managing outbound requests

Like inbound requests, all the requests initiated by Amex have to be subjected to a series of guidelines based on the protocol shown in Table 7.5, following the International Benchmarking Code of Conduct.

Further, and to help avoid legal problems that could arise from the misuse or wrongful distribution of confidential information, Amex has developed a series of proformas of confidentiality/non-disclosure agreement;

- Company-to-company benchmarking partnership
- Group participation in benchmarking studies
- Individual participation in benchmarking studies.

Appendix 7.1 illustrates the proforma for a company-to-company benchmarking partnership. With the mottos 'Borrow shamelessly' and 'Initiate to innovate' Amex is geared towards responding to the competitive challenge and to dealing with customers' increasing expectations based on the principles of: faster, cheaper, better.

The following paragraph from AMEX's benchmarking guidelines sums up the company's real intention and serious drive to create an effective culture of benchmarking:

> The goal of all quality and benchmarking initiatives is the same: to target and achieve total performance excellence. Translated, that means to continuously improve our products, services, sales, profits cost management, customer satisfaction/retention, employee satisfaction and all other key Health of the Franchise Measure areas.

7.4 Texas Instruments

Texas Instruments uses a total quality approach which is defined through world-class standards, using the Malcolm Baldrige National Quality Award. This has enabled it to set very ambitious strategies. Texas instruments' quality approach is based on five key thrusts:

- Customer-satisfaction measurements
- Benchmarking
- Stretch goals (six Sigma)
- Team work/empowerment
- Integrated quality.

Benchmarking, being one of the main pillars is described as follows:

> Benchmarking is a quality tool that allows us to assess ourselves and our competitors. Through this knowledge, teams will develop and implement plans to achieve total customer satisfaction in all that we do.

Benchmarking is defined in a variety of ways within Texas Instruments and some of the basic common terms used to explain what it is include the following:

- Benchmarking is a process of measuring products, services, and practices against 'best in class' companies and world-class practices.
- It is a tool to identify, establish, and achieve standards of excellence.
- It is a management process that requires constant updating and the integration of competitive information, practices and performance.

To demonstrate their commitment to making benchmarking embedded in the culture of the organizations a benchmarking policy was published and disseminated widely, signed by the Chairman (Appendix 7.2).

The Texas Instruments approach to benchmarking is based on the Rank Xerox methodology, using four phases and ten steps (Figure 7.4). It illustrates the form of a process approach through a series of inputs and outputs. The guidelines produced to assist teams involved in benchmarking include:

- Information sources for identifying industry leaders
- Approach to contacting partners
- How to prepare for contact/after contact
- Receiving external contact
- Check list for hosting companies benchmarking Texas Instruments
- Success indicators for benchmarking.

Texas Instruments also include a list of Do's and Don'ts:

Do's of benchmarking

- Do know your own process before attempting to contact another company for benchmarking
- Do get management commitment before attempting to benchmark
- Do identify and involve all those who might be impacted by the project
- Do implement changes or improvements identified as a result of benchmarking
- Do keep communication flow during the project, especially with a benchmarking champion
- Do use flow charting and roadmaps for your benchmarking project
- Do use internal, competitive and functional benchmarking.

Don'ts of benchmarking

- Don't benchmark just for the sake of benchmarking

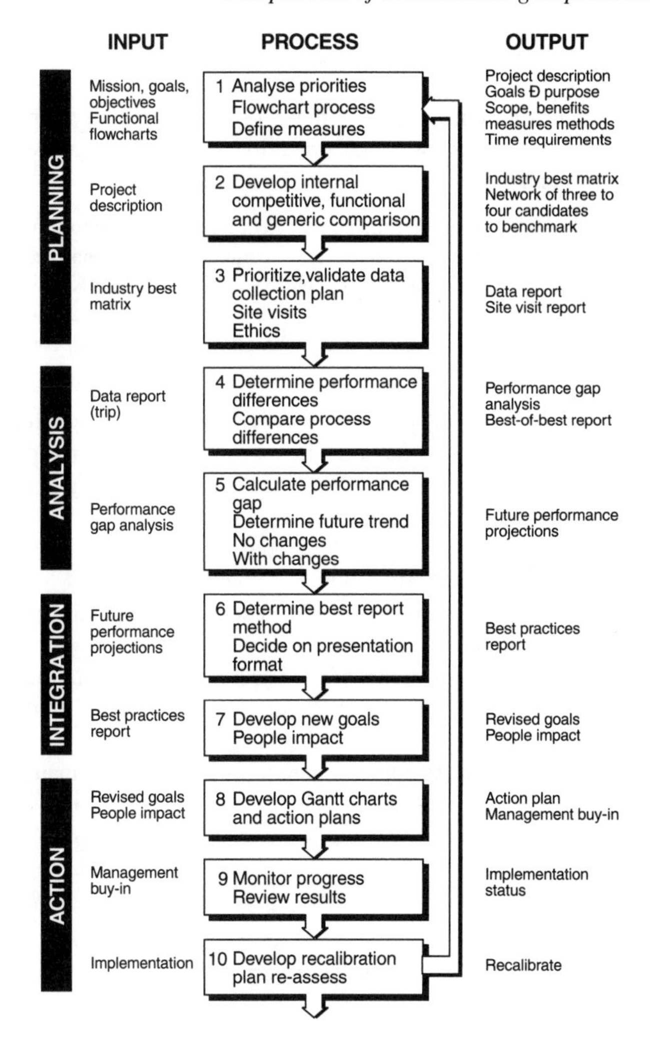

Figure 7.4 *The benchmarking process steps at Texas Instruments*

- Don't benchmark if you lack management support
- Don't benchmark unless you understand the process you are benchmarking
- Don't confuse metrics with benchmarking
- Don't confuse benchmarking with industrial tourism
- Don't ask for information you will not be willing to share
- Don't overlook public domain research
- Don't let a debrief go beyond 3 days after a visit
- Don't forget to implement.

7.5 Rover Cars

Rover uses benchmarking to ensure that world-class practices are embedded in a variety of their processes and operations. As clearly indicated in their benchmarking handbook:

> . . . we are not advocating that managers reject an opportunity to learn from an easily accessible number 3 or 4 in the world company, say a Honda pressing factory or a large Rover component suppliers. What it does mean is that this work should be undertaken with the world number 1 clearly in focus. If world class means 'Outside the Industry' then Benchmarking teams must consider this option too.

Benchmarking is used to support Rover's total quality approach, referred to as extraordinary customer satisfaction, the vision which is arrived at making them world-class. Rover, in recent times, has launched a series of initiatives to help it in its quest of achieving business excellence:

- Process capability programmes within its business units
- Fourteen strategic initiatives (e.g. Project 100, Extraordinary Customer Service, Just-in-Time)
- The creation of a process model of the business through detailing nine high-level key business processes (KBP) like 'New Product Introduction' and 'Maintenance'
- The promotion of process improvement throughout the company via the use of the Quality Action Team method.

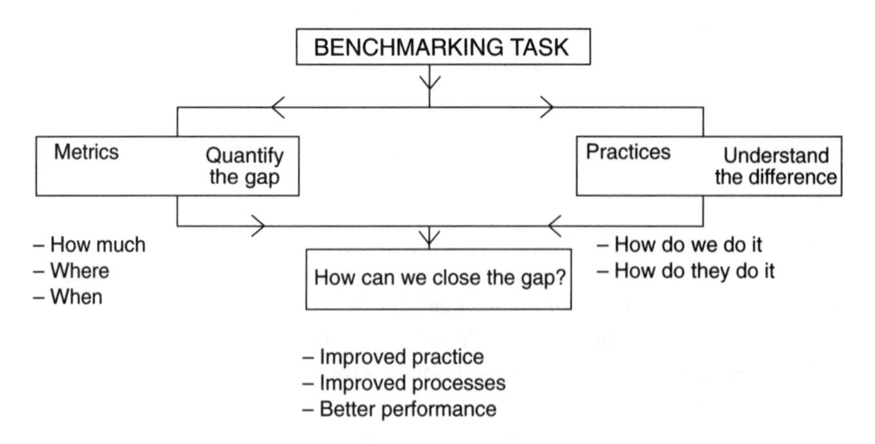

Figure 7.5 *Rover's approach to determining performance gaps*

All these initiatives are *under-primed* by benchmarking, because it puts internal performance in the perspective of practices being applied; compares against best in class and thereby helps establish action plans for closing any likely competitive gaps (Figure 7.5).

Rover uses a five-stage process based on the Rank Xerox methodology (Figure 7.6).

The Rover benchmarking approach is thorough and very detailed, similar to those previously discussed. Among the major key characteristics which are worth mentioning are:

1 The use of TQM tools to manage benchmarking projects:
   - The development of critical success factors (CSFs) – the key objectives a team must accomplish in order to achieve its mission.
   - Team building and scope definition of the project – this is in order to build the team's commitment to the value of the project, to establish the ground rules of the team, and to agree initial objectives.

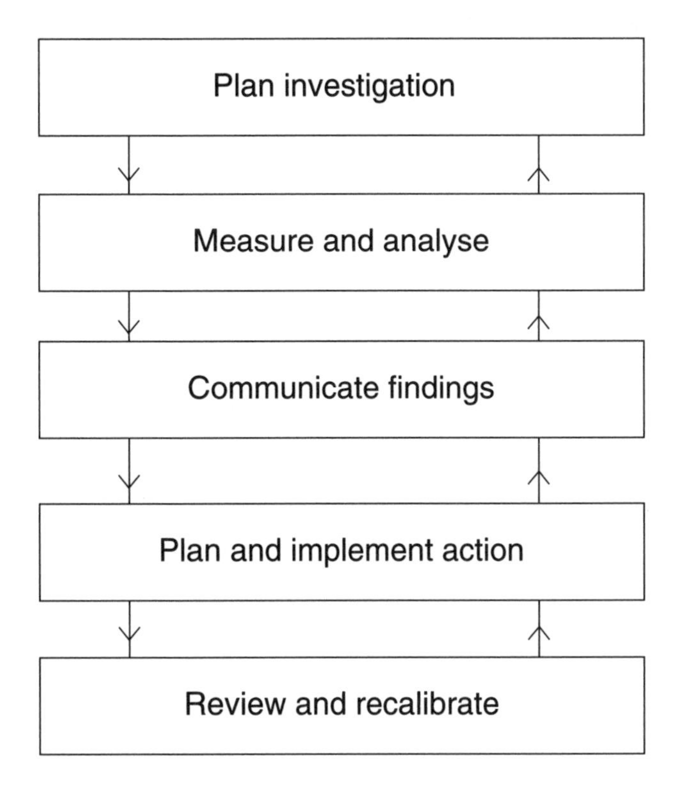

Figure 7.6 *The five key stages of competitive benchmarking at Rover*

- Use flow charting technique to define the problem accurately – to understand how the process under investigation currently operates and establish key performance measurements
- The use of solution effect diagrams.

2 The use of Imagineering – to look beyond the current method of working to generate an ideal picture of how a process might look (Imagineering is explained in Appendix 7.3).
3 Rover uses a survey to ensure that its benchmarking process is reviewed regularly and that it is going to be embedded in the culture of the organization.

Appendix 7.4 illustrates that benchmarking becomes a regular part of the way Rover operates its business operations. Team leaders are encouraged to use the questionnaire as a self-audit tool for teams involved in each specific project and also for the key stakeholders associated with benchmarking activity.

Appendix 7.1: Confidentiality/non-disclosure agreement for company-to-company benchmarking studies

American Express Company, its subsidiaries and affiliated companies ('Amexco') would like to cooperate and exchange certain information with *COMPANY* in connection with the study of comparative quality management and 'best practices', as well as the achievement of higher standards of quality in its business activities (collectively referred to as 'benchmarking'). In this Agreement, the term 'exchange' with regard to information shall include, without limitation, furnishing, providing access or availability or otherwise obtaining from any source, by any means and or through any media or means.

In connection with this exchange of information, each party may obtain or be in a position to obtain information about the other which is considered confidential, trade secret, sensitive and/or proprietary ('Confidential Information'). Recognizing that each party wishes to ensure the protection of its Confidential Information, the parties agree as follows:

1 Neither party shall be required to exchange any particular information by virtue of entering into this Agreement and, except as specified in this Agreement, no other obligations, liability or relationship shall be deemed created or modified as a result of any exchange of information hereunder. Similarly, the exchange of information between the parties shall not grant or confer or transfer any rights, license or authority in or to the information or otherwise, unless specifically agreed to in writing by the parties.
2 Both parties agree to hold in trust and confidence and regard and preserve as confidential, all Confidential Information of the other which may be exchanged as a result of the benchmarking activities as contemplated under this Agreement. In maintaining confidentiality, each party agrees that it shall not, without first obtaining prior written consent of the other, disclose or make available to any person, firm or enterprise, reproduce, transmit or use for its own benefit (except for internal purposes as contemplated hereunder) or the benefit of others, any of the other's Confidential Information.
3 The parties agree that, subject to the provisions of this Agreement, each may also disclose Confidential Information internally to its employees and employees of its subsidiaries and affiliated companies and to consultants or agents retained for purposes specifically related to that party's evaluation and internal use of such Confidential Information.
4 Confidential Information shall include, but is not limited to, all information exchanged, including, without limitation, information related to the past, present or prospective businesses, activities, operations, plans, customers, clients and/or suppliers of the parties. Information shall not be considered Confidential Information to the extent, but only to the extent,

that such information is: (a) already known to the receiving party free of any confidentiality obligation or restrictions at the time it is exchanged; (b) is or becomes publicly known or available through no wrongful act or breach of the Agreement; (c) is independently and rightfully received from a third party without restriction. In the event there is any doubt as to whether or not any particular information is 'Confidential Information', that term is used in this Agreement, the parties agree that it shall be considered Confidential Information unless and until the party which has furnished the information has confirmed to the other party that it is not.

5 Each party agrees to ensure, by agreement, instruction or otherwise, compliance with the confidentiality obligations by its employees, agents, consultants or others who are permitted access or use of Confidential Information of the other party. Each party further acknowledges and agrees that in the event of a breach or threatened breach of any provisions of this Agreement, it will have no adequate remedy in money or damages and accordingly shall be entitled to an injunction; provided, however, that no specification in this Agreement of any legal or equitable remedy shall be construed as a waiver or prohibition against any other contractual, legal or equitable remedy which either party may have under this Agreement.

6 This Agreement constitutes the entire, sole and exclusive statement of the respective rights and obligations of the parties with respect to the subject matter and may not be modified, amended or deemed waived, unless done in writing and duly signed by the parties. Neither this Agreement, nor any of its rights and/or obligations may be assigned (by operation of law, or otherwise), transferred or conveyed to any other party without the written consent of the other party and any attempts to do so shall be void.

IN WITNESS WHEREOF, the parties have duly executed this Agreement on the dates set forth below:

AMERICAN EXPRESS COMPANY	(INSERT NAME OF COMPANY)
Name: ______________________ (type or print)	Name: ______________________ (type or print)
Signed: ______________________	Signed: ______________________
Title: ______________________	Title: ______________________
Date: ______________________	Date: ______________________

Appendix 7.2: Texas Instruments' benchmarking policy

Benchmarking is a quality improvement tool that enables us to measure our products, services, and practices against those of our toughest competitors or other leading companies. Once a benchmark is identified, it provides a 'stretch goal' or vision of what is possible for business process improvements.

To achieve the TI 2000 vision, we must benchmark our processes, aggressively seeking new methods and processes to adapt to the changing environment. This means that everyone from me to the newest TIer will have to take the initiative to find these 'best practices.' We must define who does the job best, quickly close the performance gap, and surpass the benchmark.

Successful benchmarking requires a thorough understanding of our business processes before any comparisons are attempted. TI has adopted a 10-step method for benchmarking that requires planning (to identify the best companies for comparison and to determine what data to collect), analysis (to determine the performance gap), integration (to set new goals), and action (to develop, implement, and measure action plans).

All Quality Improvement Teams should adopt benchmarking as a step in their continuous improvement plan. Managers will facilitate the practice of benchmarking by expecting its use in business decisions.

Just as policy deployment has become institutionalized at TI, with time, benchmarking will become a way of managing our business.

Regards,

Jerry Junkins
Chairman, President, and
Chief Executive Officer

Appendix 7.3: Imagineering

Purpose: To look beyond the current method of working to generate an 'ideal picture' of how a process might work.

Steps
1. Brainstorm and reach consensus on which activities are absolutely essential to the process, taking account of the standards implied in the project target.
2. Construct either as individuals or as a team a range of possible processes which might deliver the required standards.
3. Refer back to existing flow chart to understand what would need to change if the suggested process replaced it.
4. Agree any proposals worth following up with further analysis.

The benefit of Imagineering is that it offers the QAT the opportunity to:

- Cut through complexities of existing activities to establish what really counts in the success of the process
- Break out of existing ways of looking at a process to see new possibilities.

However, Imagineering needs to be handled carefully. QAT leaders must be alive to three particular risks:

- Rambling conversations at too theoretical a level
- Some QAT members being turned off by discussion of theory
- The QAT using such discussions as a means to avoid taking difficult decisions.

Appendix 7.4: Rover's benchmarking assessment survey

	Strongly agree	*Agree*	*Neutral*	*Disagree*	*Strongly disagree*
Understand the benchmark process					
1. The 5-step benchmarking process is understood					
2. I understand how the benchmarks for my function were developed					
Understanding best practices					
3. There is full understanding of how benchmark partners operate					
• Their practices are better and we should emulate them					
• We can do it differently and accomplish the same/better results					
Benchmarking value					
4. Benchmarking is important					
5. Benchmarked new practices are included in action plans					
6. The value of benchmarking for setting goals is recognized					
Benchmark appropriateness to target setting					
7. Targets derived from benchmarking are, in my view, realistic					
8. We must meet benchmark targets to attain superior performance					
Benchmark communications					
9. Our planning/operating principles are based on benchmark findings					
• They are reviewed throughout the organization					
• They are understood to be based on external industry practices which we must achieve					

General comments about benchmarking

10. Benchmarking can be improved by:
11. I would like to know more about (benchmarking):
12. My specific suggestions for benchmarking are:

Title and organization

My title is .... My organization is ....

8 The diversity of benchmarking: examples of best practice

Since it is a joy to have the benefit of what is good, it is a greater one to experience what is better

Goethe

What is food to one is to others bitter poison

Lucretius

Beauty is in the eye of the beholder

Margaret Hungerford

8.1 Benchmarking in the environment

8.1.1 Introduction

Focus on the environment is becoming a growing practice among business organizations in all parts of the world. There is more pressure on companies now to develop and implement environmental policies and procedures and to start taking the impact of the environmental issues as a critical element of the competitive equation.

Various surveys at corporate level have indicated how important environmental issues are to business future success. A report by McKinsey [1] was published following a worldwide survey of top executives in 400 companies, asking their views on the environmental challenge. Three key findings were highlighted in the report:

1 More than 50 per cent of CEOs interviewed admitted that environmental issues are a major business issue for the 1990s.
2 At senior management level, there was considerable doubt as to the feasibility of dealing with environmental issues effectively and yet remain competitive.
3 Over 50 per cent of the companies analysed were primarily focusing on compliance with the law and trying to prevent incidents from happening.

Another survey carried out by Booz-Allen and Hamilton in 1991 of 200 senior executives to find out how major companies understand and manage environmental issues, has found that [2]:

- 67 per cent of executives recognize environmental issues to be extremely important to their business. This is three times the level reported in an earlier survey conducted in 1989.
- Only 7 per cent felt very comfortable and confident that their companies have comprehensive risk management systems and can deal with major risks.
- Nearly all executives interviewed have admitted that they must manage environmental risk better in the future.
- The surveyed companies have also recognized that they must take the lead by moving away from a regulatory-driven management mode (reactive/compliance) to a proactive mode, strategically driven for establishing competitive advantage.

8.1.2 The environmental issues – what are they?

- **Greenhouse effect**: Artificial warning of the atmosphere by man-made emissions: (CO_2 accounts for approximately 50 per cent of the greenhouse effect. Globally, fossil fuel combustion leads to 5.7 billion tons of CO_2 emissions per annum).
- **Depletion of the ozone layer**: The ozone layer is situated 12–30 miles above the Earth. It absorbs 99 per cent of UV. CFCs account for 90 per cent of the damage to the ozone layer. CFCs to be phased out by the year 2000 in fifty-nine countries.
- **Acid rain**: Acid precipitation damages leaves (which filter acid deposits from the atmosphere), damaging woodland. Acid emissions include SO_X and NO_X gases.
- **Habitat destruction**: Rain forest constitutes 8 per cent of land area, containing 50 per cent of all known species and anchors climatic conditions. The rain forest is lost at a rate of 200 000 km^2 per annum. It is estimated that one to fifty species per day are extinct.
- **Waste**: 1 billion tonnes of waste produced worldwide each year.
- **Transport**: 500 million cars in 1988. Motor vehicles account for 80 per cent of lead emissions, 85 per cent carbon monoxide among others.

8.1.3 What are the key challenges facing business organizations

- Awareness of major benefits to be derived from focusing on the environment and the pressures of social responsibilities makes environmental issues a corporate debate.
- Concern for public image, rising insurance premiums, increased community awareness of the environment, the implementation of stiff penalties and fines for non-compliance with environmental legislation [3].
- The emergence and growth in green consumerism. This is together with

changes taking place in consumer behaviour putting pressure on producers of goods and services to use ecologically friendly processes and ingredients for the production of ecologically friendly products and services.
- There are increasing demands on companies to manage processes, measure their performance on environmental issues, generate data and publish information on their performance. The growing pressures come from stakeholders with various interests.

8.1.4 Benchmarking in the environment – where is the starting point?

Against this background of developments it is very clear that the environment is increasingly going to become the major discussion point in boardrooms and certainly its management both strategically and operationally would create competitive advantages. A new standard for managing the environment has been developed (BS 7750) and many organizations have taken a keen interest in this. Environmental auditing is a critically important activity and this aspect of performance measurement is very often used for drawing together action plans and for the planning and development of effective strategies.

8.1.5 Booz-Allen benchmarking study

This study was conducted in 1992 [2] to gain some understanding on how environmental management relates to corporate strategy. Eight large companies were targeted for this study, for their excellence and leadership on environmental management as recognized by various bodies such as:

- Council on Environmental Quality
- World Environment Center
- Global Environmental Management Initiative (GEMI).

The study was also aimed at capturing information on practices these leading companies use in ensuring successful management of environmental issues. The sample included:

- AT&T
- Chevron
- *The Los Angeles Times*
- McDonald's
- Pacific Gas & Electric Company
- 3M
- Rohm and Haas
- IBM.

Among the top key findings are the following:

- Pacific Gas & Electric (with plants) focuses on air emissions.
- McDonald's (without plants) focuses on solid waste, particularly packaging.
- 3M focuses on NPD and the environment, for sustainable development and future regulations that might affect their product lines.
- Rohm and Haas, in the manufacture of polymers, are concerned by air, water and solid waste pollution and are among the top leaders in preventing groundwater contamination.
- *The Los Angeles Times* is benefiting from using more than 80 per cent recycled newsprint and saved a lot of money by cutting down on emissions through switching its delivery fleet to propane vehicles.
- AT&T saves money through promoting alternative materials to CFCs used widely in electronics and has achieved 100 per cent elimination of CFCs at six of its plants.

The participants were then benchmarked using the model in Figure 8.1 [2] which represents best practices in environmental leadership. All the companies involved confirmed the list in Figure 8.1 to be essential elements of successful environmental management.

8.1.6 Other examples in best practice in environment management

Procter & Gamble: first company to develop refillable pouches for fabric softener to respond to public's concern over solid waste. They met consumer demands by taking this initiative but also reduced product packaging. P&G has also continued innovating in this area through product combinations (detergent with bleach or fabric softener) using completely recycled plastic containers and recyclable packaging [4].

Black & Decker: takes back a product when its life runs out by assuming the responsibility to recycle the product (and the battery) for the customer [4].

Henkel: in 1960s–1970s, lake eutrophication became a serious environmental problem in Europe and the USA. One of the reasons for this was that lakes were starved of oxygen because of the use of phosphates in laundry detergents. This prompted leading innovating companies such as Colgate Palmolive, Henkel, Procter & Gamble and Unilever to carry out research for phosphate substitutes. By 1977, Henkel could claim that it could remove 50 per cent of the phosphates using a substitute called Zeolite. By 1986 it started to launch phosphate-free detergents and this pioneering work led to detergents containing phosphates to be phased out in Germany in 1989. Other detergent producers followed the same example [5].

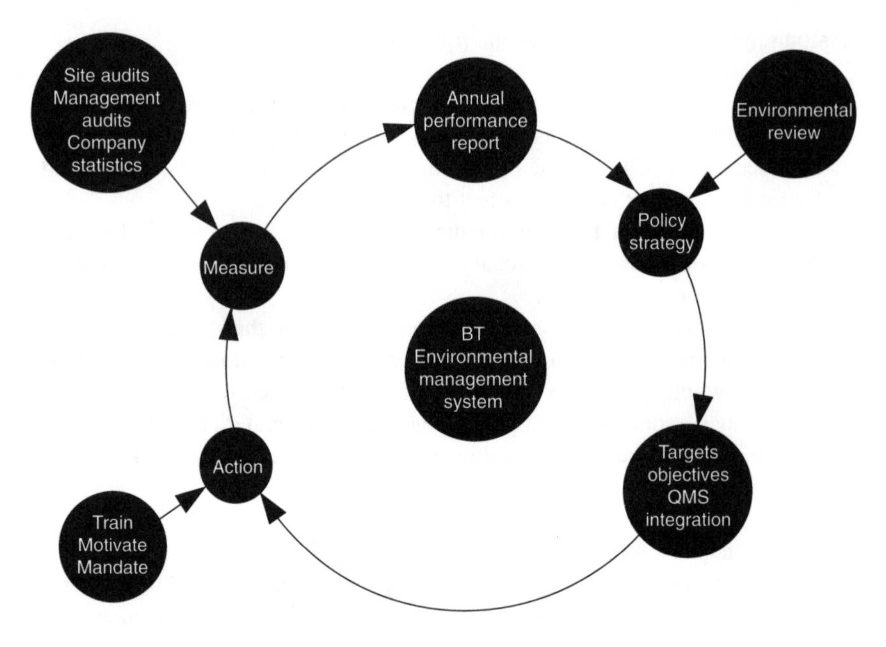

Figure 8.1 *A leader's environmental management system: the case of BT*

Volkswagen: pioneered the concept of expanded product life cycle programme to go beyond sales and service and towards total remanufacturing. With its 3V policy (Vermeiden, Verringen und Verworten) meaning Prevention (reducing solvent emissions by switching to water-based paints), Reduction and Recycling (a target of 100 per cent), Volkswagen is showing best practice in this field [6].

8.1.7 Environmental leadership role of transnational corporations

Many transnational companies are taking initiative in helping to implement environmental management programmes similar to those introduced in the West:

- AT&T seconded a senior environmental manager to China, Hungary and Russia to advise their governments on CFC alternatives.
- Chevron is leading a petrochemical initiative to share environmental know-how with Nigeria [7].
- Northern Telecom is helping the Mexican Government with technical and managerial know-how regarding non-CFC cleaning processes in electronics manufacturing.
- Union Carbide is helping Thailand in hazardous waste treatment.

- 3M is training managers from Czechoslovakia, Hungary, Poland and Turkey on environmental management.

8.1.8 Managing the environment for excellence – an integrated approach

In 1991, The Chartered Association of Certified Accountants (CACA) established the Environmental Reporting Awards Scheme (ERAS) to identify best practice in environmental management and to encourage development, implementation and the sharing of learning among industrial companies. The winners of 1991 included British Airways and Norsk-Hydro who were found to excel in the following ways [8]:

- A systematic and thorough review of core business activities
- A balanced view of environmental performance
- Independent attestation of the environmental report
- Widespread distribution of the report
- A commitment to carry on publishing further environmental reports in the future.

The criteria used for establishing the benchmarking exercise are illustrated in Table 8.1.

The winners of the 1992 ERAS Award were British Telecom (BT), the UK's principal supplier of telecommunications services, handling an average of 90 million telephone calls a day. During the financial year ending March 1992, BT had a turnover of £13 337 million, employed 210 500 people, invested £1895 million in the installation of digital exchanges and transmission technologies and spent £240 million on R&D.

The environmental management system illustrated in Figure 8.1 reflects how BT management is committed to environmental achievements. The environment is at the heart of policy/strategy development, as reported by BT's Deputy Chairman, Mike Bett: 'Our overall target for the future is for environmental considerations to play an increasingly important part in the company's planning and management process.'

The model developed by BT is based on BS 7750 and BT is working towards BS 7750 recognition. It is committed to integrate its environmental management system (EMS) into its quality management system. The key target for BT is to have all main functional units carry out environmental assessments and audits on all their activities (see Table 8.2).

Table 8.1 *A benchmarking framework for measuring high performance in environmental management*

	Key elements	*Explanation*
Set vision	Clear vision	Provides the guiding principles and policy for all environmental actions
	Corporate strategy and programmes	Delineates the means by which the environmental vision will be achieved
Design for excellence	Planning processes	Integrates environmental issues into all planning processes (i.e., investments, marketing, R&D)
	Organizational structure and responsibilities	Supports effective communication and matches environmental goals with corporate culture
Achieve continuous improvement	Performance measures	Focuses management and employee efforts to achieve the environmental goals
	Reward and recognition	Individual and team rewards explicitly linked to accomplishing performance measures
	Strategic programme	Focuses efforts on high-priority programmes such as risk management, legislative and regulatory proactivity and contractor and supplier involvement
	Training and management development	Instils the skills required to fulfil assignment responsibilities and increase environmental awareness
	Communication and information management	Ensures accessibility to relevant data and enhances decision-making capabilities
	Change management	Addresses the internal obstacles to implementation

8.2 Benchmarking quality systems

8.2.1 The role of quality systems

Quality systems such as BS 5750 and its equivalents (international ISO 9000 series and the European Standard EN 29000) play a significant role in securing markets for organizations by building in capability to deliver quality products and services, creating consistency and achieving goal congruence through the establishment of customer-supplier chains. Processes are operated and

Table 8.2 *Benchmarking critical factors for determining superior environmental management*

Independence/independent attestation
Emphasis on core business
Systematic approach to reporting and to review of issues – e.g. energy, waste, emissions, etc.
A balanced (true and fair) view is presented
Use of auditable data of actual and proposed performance
Comparative data
The report is understandable
Relationship with legislation (present and future) explored – prosecutions, etc. reported
Environmental report is available via the Financial Statements/Annual Report
Statement of policy, statement of targets, statement of performance against targets
Integration of financial and environmental matters
Relevant financial data (e.g. on provisions) is reported
Statement of how policy influences investment and other strategic business decisions
Implications of sustainability explored

managed through series of procedures. The mechanism for operating quality systems is to:

- Write down what is done
- Do what you say is carried out
- Record what is done.

Organizations can choose which parts of quality systems such as BS 5750 they apply for:

1 Part I includes twenty operating procedures with the inclusion of design and servicing
2 Part II has eighteen operating procedures excluding design and services.

Quality systems bring a wide variety of benefits, including:

- Improved utilization of materials
- Waste reduction
- Speed reduction
- Delivery reliability increases
- Increases in efficiency
- Supplier partnerships
- Customer satisfaction.

Management systems only become dynamic and lead to quality improvements through the use of quality audits and management reviews. Quality audits are a requirement of the system and the following have to take place:

- On-going review of the system at suitable intervals by management
- Appropriate corrective and remedial action taken when deficiencies are identified.

8.2.2 Benchmarking quality systems: the PERA international survey

A survey was conducted by PERA International in 1992 to assess implementation and benefits of BS 5750 [9]: A series of telephone interviews based around a sample of 2317 organizations was undertaken:

- It was found that, on average, it takes eighteen months to achieve registration.
- As a result of registration, 40 000 firms will achieve improvements in operational efficiency and 30 000 will manage to improve their marketing. Others believe that by having BS 5750 they have managed to increase their profitability.

The PERA international survey led to the following key conclusions, among others:

- 89 per cent of surveyed companies believed that the introduction of a formal management system has had a positive effect on operating efficiency
- 48 per cent of companies claimed improved profitability
- 76 per cent claimed improved marketing
- 26 per cent claimed improved export sales.

8.2.3 Benchmarking quality audits

There are two requirements of quality systems:

1. Internal audits to ensure that the system and the standards conform and that the practices in place comply with the documented procedures
2. External audits by an accredited assessor. This is in order to add value to the audited company rather than simply using the audit as a control tool.

Companies that achieve registration use the audits as a means not only for nonconformance detection but also as tools to carry out continuous improvement in all the processes and functions.

A benchmarking study was recently conducted at Bradford University in the

UK [10] to assess the use of internal audits in a small number of leading companies with the view of determining best practice. The project was also intended to analyse the interface between internal and external audits and recommend best practice. Five companies were used in this project.

Key findings

(a) **Internal audits:** Used for non-conformance purposes rather than continuous improvement. The non-conformance orientation is due to the fact that:

- Types of audits performed are functional and follow the standards rather than trail products, and processes [10].
- Lack of leadership: non-conformance audits are an easy option. Although auditors were found to enjoy the same training, it appears that in some sites people are process-oriented and others seem to be much more concerned with non-conformances and deviation from the standards.
- Variation in the efficiency of audits was found to occur between sites. Perhaps internal benchmarking would be extremely useful in helping the sharing of best practice and also in optimizing and standardizing methods and results;
- Effectiveness of the quality of audits themselves. Audits tend to be used for rectifying short-term problems without necessarily trying to implement prevention programmes with long-term solutions.

(b) **External audits:** The benchmarking study found that external audits tended to have many shortcomings such as:

- Lack of preparation for the audits
- High degree of variation in the quality of the auditors
- 'Tick in the box' audits
- No follow-up of action or feedback.

External auditors do not communicate effectively with the audited companies and a large proportion of the time is spent examining the reports from previous audits. Auditors worsen the situation by changing almost every audit, hence increasing the degree of variability of the output. There is an information gap and there has to be a way of closing it. Table 8.3 illustrates the benchmarking summary of ISO 9000 audits in the companies examined.

Table 8.3 *Benchmarking of ISO 9000 audits*

	A	*B*	*C*	*D*	*E*
% people trained for auditing	3	2	(a)	10%	3.75%
Number of internal audit/dept/year	1	1	1	1	1
Purpose of audit (b)	NC&CI	NC&CI	NC	NC&CI	NC&CI
Result of audit (b)	NC	NC(&CI)	NC	NC(&CI) depends on site	NC&CI
Type of audit (c)	IS&ID	IS&ID	IS&ID	ID	IS&ID
Audit organization responsibility	Auditor	Auditor	Auditor	Auditor	Auditor
Documents produced	Report CAR	Report CAR	Report CAR	Report CAR	Report CAR
Follow-up made	Site manager	Quality manager	Site manager	Site manager	Div. manager
Level of top management involvement	Medium	Medium	High	High	Medium
% non-conformances corrected as agreed	(d)	26–73% (e)	(d)	95%	65–70% (f)
External assessor	Veritas	BSI	BSI	BSI	BSI
Number of external audits/year	4	2	2	2	2
Number of times each dept audited	2 (g)	1 (h)	1 (h)	1 (h)	1 (h)
Satisfied with external assessor	Yes	Situation improved	No	No	Situation improved

(a) This information is considered sensitive by C. Its publication has not been authorized.
(b) NC: Non-conformance.
CI: Continuous improvement.
(c) IS: Inter-site.
ID: Inter-department.
(d) No figures available.
(e) The results are fairly irregular varying over time from 26 per cent to 73 per cent.
(f) The measure is: 'corrected first time' and not 'corrected at the agreed date' because the follow-up audit may not take place until some time after the corrective action completion date.
(g) At A, every department is audited twice a year, internally and externally.
(h) In the other companies, each department or process will generally be audited once a year. It can be audited a second time, externally by BSI.

8.3 Benchmarking in the public sector

8.3.1 Introduction

Various areas of the public sector use the art of benchmarking without perhaps knowing that they do so and without necessarily exploiting the full potential from benchmarking exercises in order to strengthen processes and continuously introducing best practice. Local authorities, for instance, compile trends of performance and compare them against national average. Good results or disappointing ones are then explained by referring back to the processes concerned. In more recent years local authorities started to incorporate European trends as a result of the common market among European countries.

An example of benchmarking application in the public sector is the Highways Services. Each local authority is committed to reducing highway and transportation accidents through investment and various programmes geared towards minimizing or eradicating accidents of any kind.

8.3.2 Benchmarking in Highways Services

The role of Highways Services is to ensure that accidents of all kinds do not happen. As expressed by the following statement from one of the local authorities in the UK: '... Council will do everything in its power to reduce accidents, but road users are ultimately responsible for the greatest reduction by adjusting their driving techniques to suit the road conditions.' Accidents covered are all those which result in personal injury. The following is a classification of types of accidents:

Casualty: A person killed or injured in an accident. One accident can result in more than one casualty. Injury could be classified as slight or serious

Fatality: A person killed during or dying within 30 days of an accident

Serious injury: An injury for which a person is detained in hospital (e.g. fractures, concussion, internal injuries, crushing)

Slight injury: An injury of a minor nature such as a sprain, bruise or cut which is not judged to be severe (e.g. slight shock)

Accident severity: Severity of the most severely injured casualty (fatal, serious or slight).

The benchmark set by the UK Government is to reduce accidents by **one-third** by the year 2000. On average, it is found that each accident produces 1.3 casualties. In order to achieve the target set by the government, various schemes were set up to help reduce accidents:

1 Schemes aimed at educating and changing drivers' attitudes through better awareness, provision of information and appreciation of hazardous situations.
2 The development of appropriate legislation. For instance, seat belt legislation was found to have made a contribution in reducing the number of casualties from accidents. In addition, stricter driving standards for bus and coach drivers are leading to lower casualty records.

There is also economic implications for maintaining accident rates at a low level. It is thought, for instance, that the average cost per accident is around £26 000. In one of the district communities, the cost of accidents in 1992 amounted to £37.5 million.

Figure 8.2 illustrates casualty trends for the period 1985–1992 benchmarking performance at district, regional and national levels. Figure 8.3 illustrates

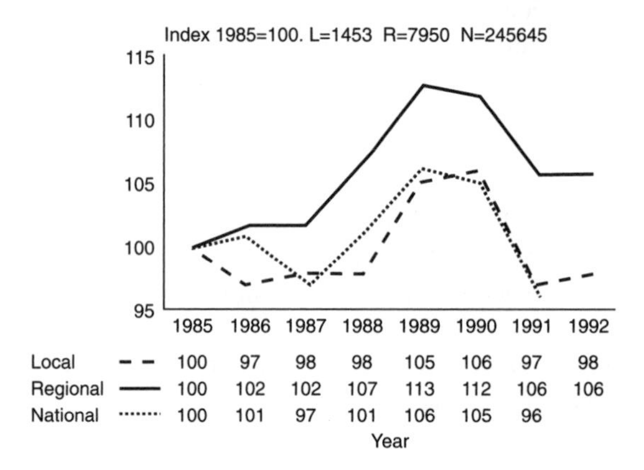

Figure 8.2 *Casualty trends benchmarking performance 1985–92*

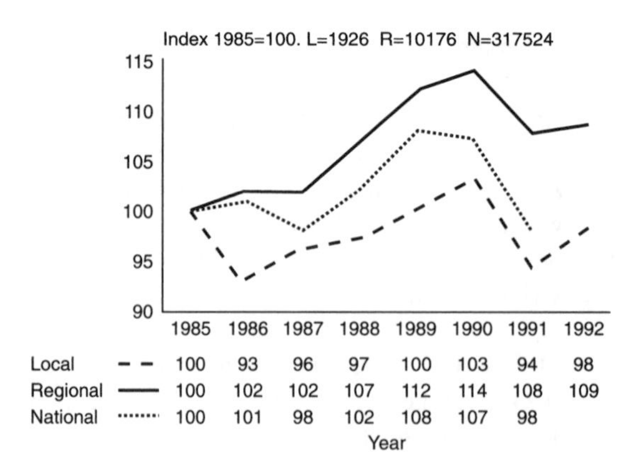

Figure 8.3 *Injury trends benchmarking performance 1985–92*

injury trends for the period 1985–1992 at district, regional and national levels.

Causes of accidents are most often found to be directly attributable to human error (95 per cent). The following are examples of human error:

- Pedestrian crossing
- Misjudged speed
- Going too fast
- Driving too close
- Overtaking
- Alcohol/drugs.

A small number of accidents is found to result from factors such as environmental or vehicle defect factors. These include, for example:

- Falling objects from vehicles
- Sudden mechanical failures
- Obscured vision
- Road environment.

Table 8.3 illustrates a comparison of casualties by group. The statistics show a five-year average trend, and whether it is on upward or downward slopes, and the percentage change from previous year. Figure 8.4 illustrates casualty trends over 1985–1992 between various groups.

8.4 Benchmarking in the utilities sector

The utilities sector has been chosen as a specific industry because, unlike other industrial sectors, it is faced with so many different challenges. This industry, although in recent years has been more geared towards greater efficiencies and being increasingly competitive, still places more emphasis on safety first, unlike any other industrial sector. In addition, this sector is highly regulated and has to comply with stringent regulations and must meet continuously set standards. Before examining in specific terms one or two examples of companies that have implemented the art of benchmarking and who operate in the utilities sector, first let us try to deal with the question of whether benchmarking is an alien concept to utilities.

Benchmarking in the nuclear industry is a well-established concept, although it has not been referred to in those specific terms in the past. One can go back as far as 1979 when the Institute of Nuclear Power Operators (INPO) was established in the USA to promote the highest levels of safety and reliability. In other words, the INPO was established to promote excellence

Table 8.4 *A comparison of casualty per group: benchmarking performance over 5-year plan*

	1985	*1986*	*1987*	*1988*	*1989*	*1990*	*1991*	*1992*	*5-year ave. 85–9*	*Percentage change on 5-year ave.*	*Percentage change on 1991*
Pedestrians	444	436	429	442	426	481	422	405	435	Down 7%	Down 4%
Cyclists	84	98	97	97	126	100	113	118	100	Up 18%	Up 4%
Motorcyclists (inc. pillion)	345	378	279	264	244	238	165	151	302	Down 50%	Down 9%
Car drivers	483	435	490	529	585	617	617	630	504	Up 25%	Up 2%
Car passengers	424	339	383	388	432	423	380	423	393	Up 7%	Up 11%
Others	146	121	166	141	118	131	123	152	138	Up 10%	Up 23%

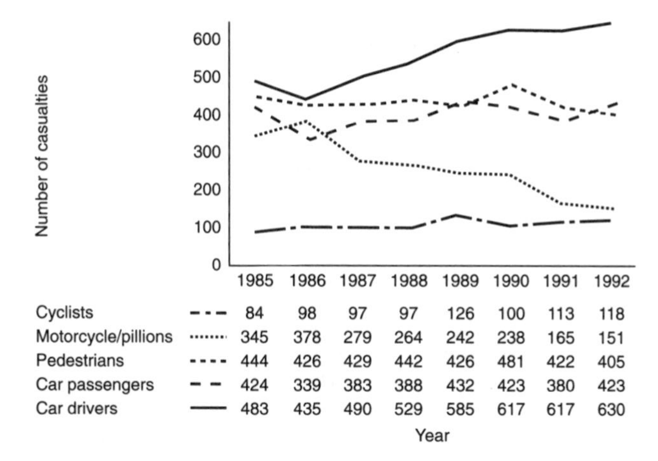

		1985	1986	1987	1988	1989	1990	1991	1992
Cyclists	– · –	84	98	97	97	126	100	113	118
Motorcycle/pillions	········	345	378	279	264	242	238	165	151
Pedestrians	- - - -	444	426	429	442	426	481	422	405
Car passengers	– –	424	339	383	388	432	423	380	423
Car drivers	——	483	435	490	529	585	617	617	630

Figure 8.4 *Casualty trends benchmarking performance per group 1985–92*

in the operation of nuclear power plants. In those days the term 'benchmarking' was not used specifically, but the mission of INPO was very much to promote best practice and to ensure that standards of safety are increasingly tightened.

- For example, the use of expertise and experience from within the nuclear community is promoted in order to provide industry with best methods, best technologies and best practices.
- INPO was also established to promote and encourage the exchange of information on good practices and lessons learned from all the nuclear operating companies, recognizing that by improving and refining existing practices, safety standards can be raised and disasters can be avoided.
- INPO encouraged the involvement and participation of personnel from different functions within utility companies to promote and exchange information and to enhance the knowledge levels of all people involved.

Utilities companies from at least fourteen different countries joined the INPO corporate membership, and one could argue that since its creation INPO has helped to enhance standards of safety worldwide to an excellent level and the fact that there are very few disasters taking place is perhaps due to the reason of encouraging best practice, problem solving, exchange and cooperation and networking between the various operators worldwide.

Followed by INPO in 1989 the World Association of Nuclear Operators (WANO) was created. Based on the similar principles of INPO, WANO is used to facilitate the exchange of information between various nuclear plant operators worldwide, and to encourage comparison, emulation, communication, exchanges, partnerships and joint projects between the various corporate members. The objectives of WANO are as follows:

- Operating experience exchange
- Operator-to-operator exchange
- Plant performance indicators
- Sharing of good practices.

WANO has ten performance indicators used for benchmarking. All the data are compiled by WANO and made accessible to all the corporate members for learning from each other. In addition to INPO and WANO, various operators in utilities participate actively in the International Atomic Energy Agency (IAEA). The IAEA is an ideal forum for sharing standards of excellence and learning high standards of safety and reliability, as well as operating nuclear power plants in an economic and efficient manner.

It appears, therefore, that benchmarking in utilities is not an alien concept and has been applied for a large number of years in various ways. The utilities sector is an interesting one to look at from the point of view of benchmarking, particularly of the recent changes that have taken place, particularly in the UK and the restructuring that has happened within the Central Electricity Generating Board (CEGB). While previously CEGB as one organization used to enjoy a monopoly position in the bulk supply of electricity within England and Wales, now its restructuring has led to the birth of various operators including, for example, Nuclear Electric and National Power. They have now to operate independently from each other, and compete in the private electricity market, and this commercial challenge will determine how successful each is going to be.

8.4.1 Benchmarking at Nuclear Electric

Nuclear Electric has used the art of benchmarking to enable it to develop stretch objectives and put its vision in place. Its vision is 'We will be a quality company, we will make an operating profit before the levy in 1995, and be a key part of the country's energy supply into the twenty-first century'. Through benchmarking, Nuclear Electric has established its strategic framework for driving for the desired goals and has established some performance goals in the following areas:

- Manageable costs per unit of production
- Numbers of staff
- Output per employee.

Internal comparisons

Nuclear Electric has been using the art of benchmarking for both internal comparisons and external comparisons:

1 **Plant evaluation**: An evaluation has taken place using a group of peers for the purpose of identifying good practices and comparing them against internationally accepted standards. Internal benchmarking within Nuclear Electric is really for identifying areas for improvement in all the activities, particularly those that affect safety and reliability standards.
2 **Investors in people**: This is an opportunity for Nuclear Electric to use its human resources more proactively and to encourage innovativeness at all levels in order to meet business objectives. The investors in people programme, initiated by the government, was used at two lead sites within Nuclear Electric, particularly to determine areas for improvement.
3 **Cost management review**: This exercise has taken place in order to look at costs in relation to activities and their worth for overall company objectives. It has helped the company to identify opportunities for improvement and proceed with plans for rationalizations, and has also enabled a comparison to take place between the various stations and the numbers of staff involved in all the various activities.

External comparisons

Various benchmarking exercises have taken place within Nuclear Electric, benchmarking its operations against other utilities and also in the wide industry community as a whole.

1 **Outage management**: The benchmarking exercise studied Nuclear Electric's average outage time in all its power plants and basically identified opportunities for reducing at that level. As a result of this benchmarking exercise, in two power plants the outage was reduced from 17 weeks to 10 weeks with a further opportunity to reduce this standard to even 8 weeks.
2 **Safety case management**: This benchmarking exercise scrutinized the whole safety case management system. The exercise is still ongoing and the purpose of doing it in the first place was to simplify the structure and get rid of the complexity while maintaining the essential elements of safety case management.
3 **Health physics**: This is a benchmarking exercise with all the utilities in order to establish best practice in radiation protection activities.

Benchmarking within Nuclear Electric has really worked. It has given it a new lease of life since vesting day and there is evidence to show this. Since vesting, output has gone up, costs have come down, and safety standards have been maintained and even improved. In terms of achievement as a result of benchmarking, the following have taken place:

- Output increased 29 per cent
- Market share increased 26 per cent
- Costs were reduced by 22 per cent
- Productivity increased by 53 per cent.

The journey still continues at Nuclear Electric and its target for 1995 was to make a profit by re-examining its manpower, flexibility and utilization and bring about the necessary changes in a more radical way by assessing its phase costs and focusing on substantial productivity increases. Benchmarking has given Nuclear Electric the confidence to establish stretch objectives and aim for achieving them. Striving for excellence in Nuclear Electric is becoming a way of life and in the words of its chief executive Dr R. Hawley, this is how Nuclear Electric is striving for competitiveness [11]: 'In the private electricity market, there is no room for other than the excellent, our competitors will see to that.'

8.4.2 The case of Florida Power and Light

Florida Power and Light was established in 1925. It is the fourth largest investor-owned electric utility in the USA, and certainly the fastest growing in terms of numbers of customer accounts. Based on 1989 figures, its serviced territory spreads to a population of approximately 5.7 million, covering 27 650 square miles, which is approximately one-half the state of Florida. Florida Power and Light, based on 1989 figures, was employing 15 000 employees, had facilities including thirteen operating plants, seven operating offices, forty-five customer service offices, seventy-two service centres, 397 substations, 53 300 miles of transmission and distribution lines and, more importantly, 3 million customer accounts. Florida Power and Light started experience with quality back in the 1970s as a result of the various initiatives introduced in nuclear power plant construction, and many achievements were registered, for example cost containment and timing were improved and quality control and public safety measures were more stringent. However, it wasn't until the early 1980s when quality was introduced systematically at a corporate level using the expertise of gurus such as Joseph Juran, Edward Deming and Philip Crosby, among others. The foundation for building quality into Florida Power and Light and those which led it to win the Deming Prize, the most prestigious quality award, in 1989 were the following principles:

1 **Customer satisfaction**: The need and reasonable expectation of customers must be satisfied.

2 **PDCA (Plan–Do–Check–Act)**: Continuous improvement based on working at the process, planning what we do, carrying it out, checking what we have done, acting to prevent error and improving the process rather than repeating the PDCA cycle time and time again.
3 **Management by fact**: This concept has two aspects to it: first, collecting data objectively, then using the refined information to solve problems and acting in order to improve quality in all aspects.
4 **Respect for people**: This is fundamental to the Florida Power and Light philosophy in that all employees are self-motivated and encouraged to use their creative output. Their skills and knowledge are continuously harnessed to ensure that there is optimum utilization and, most of all, valuing people as an integral part of the organization and having respect for them.

Through quality, Florida Power and Light has achieved significant results, some of which are not measurable. For example, it has developed a new corporate culture where all employees have a renewed sense of pride, respect and satisfaction. The climate of work is based on multi-functional levels and where everybody is given the opportunity to contribute and use their creative talent.

Some employees have started to use the continuous improvement approach on their everyday work and opinions of Florida Power and Light from outside have changed significantly better because of quality achievements beyond the customer and the media and other stockholders, government agencies, politicians, and even the regulators, have started to admire the way Florida Power and Light was driving its business towards excellence.

Benchmarking at Florida Power and Light

When asked the question 'How will you know when you are the best managed electric utility company and how will you measure it?' the previous Chief Executive of Florida Power and Light, Mr Hudiburg answered as follows: 'At FPL we will learn from anyone and we will borrow a good idea from anyone.' Based on the Rank Xerox methodology in 1988, Florida Power and Light developed their own approach to benchmarking, first by selecting fifteen other leading electricity utilities to compare them against Florida Power and Light using information on size, reputation and location. Then they tried to obtain data on the various functions, particularly data that could be used to compare quality indicators. The information obtained was useful and included the following areas: changing price; lost time injuries; nuclear regulatory commission (NRC) violations; service and availability; public service commission (PSC) complaints. Going back to figures in 1985 when Florida Power and Light did not measure up at all against the fifteen top performers in utilities, the figures obtained in 1988 seemed to indicate that they moved from thirteenth to third place and by 1989 they were moving towards second place. As

Figure 8.5 demonstrates, Florida Power and Light achieved great improvements in the various performance measures examined, mainly because of the quality drive and its commitment to achieve high standards of performance in all aspects. From 1986 to 1989 the percentage of Florida Power and Light customers who reported that they were extremely satisfied had increased from 45 per cent to 55 per cent.

Florida Power and Light has used benchmarking by referring to quality indicators – those that measure the important things to the customers. It has learnt from the onset to start with the customer and has subsequently changed all its processes to focus on the customer continuously. Through the inspiration they have from the Japanese, Florida Power and Light have learnt that by focusing on the customer, the bottom-line results will be enhanced significantly. As explained by their previous chief executive [12]:

> The Japanese study the customer. They then study the customer again. Their entire process, internal as well as external, is directed toward the customer. By serving the customer they believe they will gain market share and corporate health. It is taken as given that if they are better than anyone else at serving the customer, they will have strong savings and thereby make greater profits.

Customer service used in utilities tend to be very similar between the various operators. For instance, the customer survey conducted by Florida Power

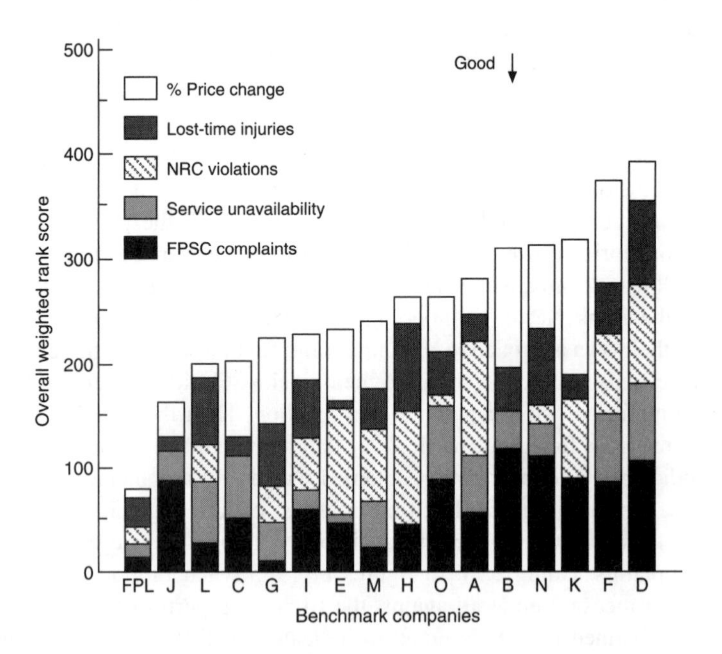

Figure 8.5 *Benchmarking in utilities: the case of Florida Power and Light*

and Light in 1986 to identify programme areas and customer needs, included the following measures:

- Accurate answers
- Timely action
- Accurate bills
- Considerate customer service
- Customer programmes and services
- Continuity of service
- Understandable rates and bills.

This was very similar to one conducted by a UK-based electricity services supplier who carried out the survey with the same objective, i.e. in order to identify problem areas and customer needs. The measures used were the following:

- Supply reliability
- Price
- Speedy response to telephone calls
- Speedy response to correspondence
- Reliability in keeping appointments
- Making appointments designed to meet customer needs
- Approachability/helpfulness of staff (particularly in shops)
- Fair treatment of customers
- Accuracy/understanding of accounts.

8.4.3 Benchmarking – the regulators' perspective

Benchmarking utilities is a very interesting subject to study because it is a very highly regulated industry. The role of the regulating bodies is to protect customer and consumer interests as far as price is concerned as well as quality of service. The role of the regulators is to make sure that customers and consumers get value for money all the time. They are empowered to monitor levels of service and also to set standards which cover aspects such as reliability or product and services supplied, the availability of what is supplied and the services surrounding the customer and company interface, i.e. those services which deal with customer complaints, queries and involvement.

Regulators focus on output rather than input. As such, they concentrate on the services and quality of the products supplied, rather than the processes which deliver those and the way they are managed by organizations. They also ensure that customers are involved and that suppliers are addressing their needs efficiently and involving them in the various decisions. The regulators are not specifically interested in individual companies, but cover the industry as a whole. One of the key areas where regulators become involved is the availability of information and its transparency. The information obtained enables the regulator to conduct benchmarking exercises in order to identify competitive parameters and to establish standards, so therefore the key activities that they will focus on include:

- The measurement of service delivered to the customer
- Publishing the results
- Rewarding good performance.

The regulators, therefore, are not interested in what is measurable and what companies publish in terms of the performance measures. They are more concerned with measurement from the point of view of the customer.

This transparency of information and its availability is perceived to be very important. Indeed it is considered that the measurement of performance in itself is not enough. The results have to be made available to the individual companies concerned as well as for customers with a view to achieving the following:

1 The customers are entitled to know how the quality of the service they receive from particular companies compares with that provided by others, and thus enables them to arrive at decisions of future involvement with their suppliers.
2 The individual companies themselves need to know where they are in relation to other providers and for them to work hard at closing the gap. It is an opportunity for them to identify best practice and incorporate it into their internal operations so that their standards match those of best in class.

This is perhaps the direct way in which regulators can apply the art of benchmarking and can influence individual organizations in closing gaps and achieving high standards of performance from delivering value to the customer to achieving high quality. Florida Power and Light, for instance, have found that by independently and voluntarily buying in the quality ethos and by achieving high standards and setting tighter targets for themselves, they become in a position to meet minimum standard requirements set by the regulators, but in a proactive way they can influence the regulators' decisions in setting standards because they have become a model company that other suppliers of utilities are aspiring to follow.

Benchmarking can also be applied by regulators to look at various standards from different industries as the following examples indicate [13]. Figure 8.6 illustrates results from a number of customer surveys conducted by the specific industries concerned asking customers to rank services according to the order of priority, and by compiling the various surveys, an interesting

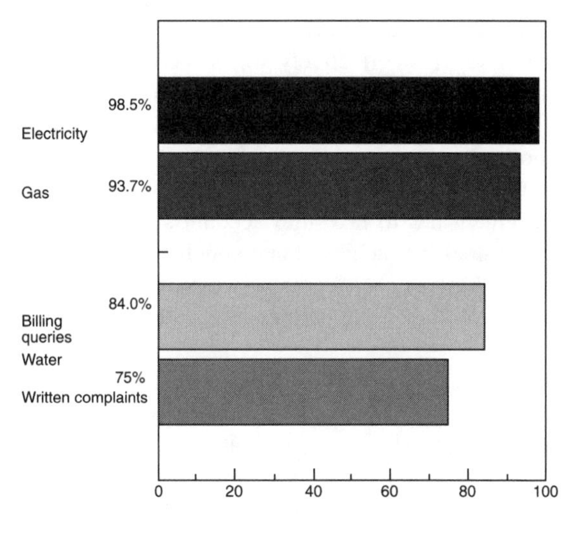

Notes:

Electricity: responses to written queries within 10 working days
Gas: responses to written queries within 5 working days
Water: billing queries – responses to telephone and written queries within 10 working days
written complaints – responses to written queries within 10 working days

Figure 8.6 *Benchmarking – the regulator perspective: provider's performance*

benchmarking exercise could take place between the four different utility suppliers involved.

Another survey commissioned by the electricity regulator [13] has confirmed the rankings of service quality identified by the earliest surveys referred to in Figure 8.6. As the figure indicates, based on the question of whether customers are satisfied or dissatisfied with the overall service provided, British Gas is top followed by a local electricity company, British Telecom, and lastly a local water company.

In conclusion, therefore, the art of benchmarking is applicable to utilities although they are a very special industry because of the regulations imposed on them and also because of the need to have high safety and reliability standards.

8.5 Best-practice supplier partnerships

8.5.1 Introduction

Managing supplier relationships in a modern business context is no longer considered a 'distraction' from doing important tasks. On the contrary, this is one of the most critical areas for superior competitiveness. For those companies that manage their suppliers effectively, this is a core competence. Supplier management in the 1990s is no longer the function of the 'buying or purchasing manager', it is a corporatewide activity which relies on a cross-functional, process-based approach where suppliers become an integral element of the value-adding chain.

Strategic management has to take into account supplier issues at various levels, to ensure that goals are achieved and supplier performance is compatible with company objectives [14]. Perhaps two issues need to be considered from the onset in developing strong partnerships with suppliers:

1 **Supplier evaluation and selection:** Supplier evaluation and selection is not merely a financial or functional consideration. Very often organizations consider suppliers which can be 'bolted on' to their existing culture or, alternatively, the option is given to suppliers who are more likely to consider changing and radically re-engineering their processes to fit in their customer requirements. Visionary organizations, however, may even consider internal changes and internal re-engineering in order to capture the full potential from their relationships with their key suppliers. Unlike traditional approaches to supplier evaluation, some of the models suggested cover various aspects [15]:

   - Financial issues
   - Organizational structure and strategy
   - Technology
   - Business credibility and other factors.

2 **Supplier management:** The management of suppliers in a modern business context is not 'transaction based' but process-focused. The spirit of continuous improvement has to be extended to suppliers and the optimization of internal processes linked to the maximization of external processes such as supplier management. Cost reduction should not overshadow wider benefits in the areas of quality improvement, responsiveness, flexibility, innovation and involvement in problem solving. Some of the issues which need to be addressed during the development of close relationships with suppliers should include [16]:

   - Consistent payment of invoices
   - Long-term commitment to suppliers
   - Early involvement of suppliers
   - Consideration of suppliers' feedback
   - Supplier access to key people.

   Effective supplier management has to also be dependent upon some 'hard necessities' such as [17]:

- A clear definition of requirements
- A selection of qualified suppliers
- An agreement on quality assurance
- A provision for the settlement of disputes
- Incoming inspection control
- The receipt of quality records.

Effective supplier partnership models are only those where human bonds are established at all levels through corporate commitment, positive communications, openness and willingness to share knowledge and expertise and joint vision development. Some of the recommended approaches in facilitating the process of effective partnership development through team building, include [18]:

- Identifying the willingness to embark on a partnership-enhancing process
- Identifying mutual issues
- Selecting participants
- Selecting realistic expectations
- Fashioning ground rules
- Structuring, monitoring, and managing the problem-solving process
- Avoiding and managing conflict
- Considering an outside facilitator-consultant
- Gaining resolution and ensuring action-taking
- Learning from experience and planning the continuation.

8.5.2 Examples of best practice

The following examples provide pioneering and effective practices of supplier–customer partnerships. The models highlight represented integrated approaches, usually driven from a top-down, strategic perspective and exhibiting interactions at various levels. These case studies do also show the power of total quality management and how a commitment to continuous improvement can help forge better links. At the heart of all quality drives is the emphasis on performance measurement and action taking.

Lucas

Supplier development process

Lucas Automotive have responded to competitive challenges by changing its relationships with suppliers, in an attempt to match the best in the world. Following a successful pilot project in the firm's heavy-duty braking division, sourcing reforms were introduced at Lucas Group plants worldwide.

For European manufacturing companies generally a 1 per cent cut in spending on materials can give a financial benefit equal to an 8–10 per cent increase in sales. Companies have traditionally placed a heavy emphasis on negotiating the lowest piece price from their suppliers. A new approach was designed and emphasis was placed on quality to reduce costs throughout the sourcing process. In its drive to improve product-development lead times and quality, and cut total-acquisition costs, Lucas decided to concentrate on a limited number of very close partners. These links covered aspects of planning, engineering, manufacturing and finance which required a commitment from the most senior level.

The greatest opportunities were at the design stage where it was considered that suppliers could assist in the task of achieving the optimum design for manufacture. Suppliers were involved with the product design team as opposed to making a product from a drawing. As the suppliers' responsibilities have increased, more of Lucas's own engineering resources have been released.

The Strategic Process of Supplier Partnership Development involves four phases:

- **Phase 1: Research**: This lasts for 4–6 weeks and is carried out by interview and literature search. The main tasks are to
  - Analyse current and future spending
  - Determine the structure of the supply industry
  - Understand the technology available to suppliers
  - Build an initial model of the total acquisition cost.

  Long-term saving opportunities are also worked out.
- **Phase 2: Evaluate**: Evaluating suppliers lasts between 6 to 8 weeks and the characteristics required of a supplier will vary according to the item being bought. Suppliers are assessed according to the different technologies and skills of the suppliers against the required characteristics and a world-class benchmark.
- **Phase 3: Structure**: After establishing the implications of the partnership with those short-listed suppliers, the co-makers are selected and the supply relationship structured. The long-term nature of the relationship means that it is vital that the co-maker should achieve world-class performance levels. Benchmarking helps to define where cost-reduction efforts should be concentrated. Joint targets are set and improvement plans approved, and this process takes two to three months.
- **Phase 4: Implementation**: Where business is transferred from another supplier to the co-maker, a joint design review is conducted for each part. The improvement plans are regularly reviewed and the buyer coordinates activities by Lucas to help the supplier to improve his or her processes. Implementing the relationship and organizing for continuous improvement lasts between 6 months and a year.

Lucas is examining changes to some of its structures and skill levels, to take account of the new arrangements, but the following remains to be done:

- Strengthening Lucas's ability to help co-makers with improvement programmes
- Extending to co-makers the benefits Lucas has obtained on raw materials and services and
- Developing supplier groups to optimize product design and spread best practice.

Achievements are shown in Table 8.5

British Telecommunication Ltd

Philosophy

Major purchases are carried out centrally where an increasing number of corporate agreements are being set up. Supply opportunities also exist through local procurement units and, on a sub-contract basis, through BT's main contractors. In addition, BT has several subsidiaries and associated companies, each with their own purchasing units.

With such buying on a large scale, it is important that the company gets value for money, which is why emphasis is placed on quality, reliability and price. Quality is the cornerstone of BT's strategy. It is of equal importance to its customers who recognize that price is important but quality is crucial. As BT operates at the forefront of technical change in an increasingly competitive market, it is vital that the position of both the customers and BT is safeguarded by buying only those goods and services which represent value for money.

Table 8.5

Customer benefits	*Mutual benefits*	*Supplier benefits*
Fair Price	Equitable pricing based on open-book cost modelling	Fair margin
No double-tooling cost	Scale economics	Increased volumes
Long-term agreement	Less uncertainty	Better forecast
Less component variability due to single source	Better quality	Clearer understanding of requirements
Lower long-term cost of acquisition	Long-term perspective	Encourages investment
High support and responsiveness	Increased mutual dependence	Not forced to be a 'yes' man

For this reason, a close relationship with suppliers is essential. Such a partnership is founded on trust, cooperation, support and, above all, continuous improvement.

Supplier requirements

As the partnership develops, the supplier will need, at various stages, to:

- Show commitment to meeting the requirements of ISO 9000 and to introduce systems such as electronic data interchange and barcoding
- Discuss problems in an open and timely manner
- Participate in joint quality improvement projects
- Cooperate in comprehensive monitoring of performance
- Provide information on whole life costs
- Identify and improve any process which might harm the environment.

BT expects its suppliers to have high-quality reliable goods and services, first-class delivery to schedule and excellent support services. It prefers to buy proprietary services against international standards leaving design questions to the suppliers instead of issuing detailed product specifications.

For a successful partnership, commitment is required from both sides, and hence BT ensures that:

- The professionalism of its staff will be used to purchase goods and services in a way which meets the European Community legislation and is fair to all participants.
- It gives a better view of its forward plans so that suppliers can invest in quality improvement, product capacity and development with great confidence.
- It provides better feedback on overall performance.
- It clearly defines its requirements; to help with this it has rationalized its internal processes.
- It continues its companywide total quality management programme, supported by ISO 9000 accreditation.

Supplier development process

Initial enquiries are addressed to a central contact site. A potential supplier will be invited to provide information on the company profile, goods/service range, financial accounts and quality assurance structure. This will be studied against the requirements of BT's qualified supplier list. A vendor appraisal will be conducted prior to issuing invitations to tender.

Quality standards

BT's quality standards are applied according to the type of item to be bought and to the degree to which it affects BT's service to its customers. They are classified into sectors as follows:

- **Sector 1:** This covers low-risk purchases for internal consumption, cash transactions, vehicles and energy needs. The quality standards required are that the products conform to standard terms and conditions.
- **Sector 2:** This covers items like personal computers, computer maintenance, tools, most engineering stores, freight services, underground work, etc. In addition to Sector 1 requirements, the supplier is expected to be committed to ISO 9000 and have a proactive approach to quality.
- **Sector 3:** This covers goods and services that directly impact on the service imparted to BT customers. In addition to Sector 2 requirements, joint quality-improvement programmes and vendor ratings are used to monitor and measure supplier and product performance.
- **Sector 4:** This covers critical purchases vital to the operation of BT's core activities and encompasses switching, transmission and network management computing. Quality standards require suppliers to participate in reliability-improvement programmes to comprehensive whole-life costing studies.

BT also has additional generic standards addressing issues such as the environment, human factors, safety, etc.

Nissan

Philosophy

The advent of Japanese companies such as Nissan into the European automotive industry influenced a radical shift in customer–supplier relationships, and the traditional adversarial relationship has given way to closer cooperation. This is because manufacturers now realize the importance of the supplier's contribution to competitiveness. Competitiveness cannot be achieved in the 1990s by the OEM manufacturer alone, and more success will be gained if its suppliers are aligned to its strategic vision. The effectiveness depends upon the supplier's ability with respect to the following:

- Insight into the market, customer and competitive needs
- Flexibility and responsiveness to competitive thrusts
- Efficiency in the use of combined assets
- Elimination of any non-value added activities in the chain
- Adoption of a vigorous culture of 'continuous improvement'.

The increasing trend is to source a greater proportion of components from outside, thus reducing the amount of vertical integration, and for the manufacturer to become a designer and assembler of products. Unlike traditional European mass producers, Nissan does not retain a high proportion of the value chain in-house and is not involved in the detailed design. This means that the supplier has a very high profile at Nissan.

Nissan tend to source sub-assemblies from fewer suppliers rather than components which comprise the lower levels of the bills of materials. This enables Nissan to pay more attention to fewer suppliers. The costs that would be associated with dealing with a larger number of suppliers are eliminated and more attention can be paid to the remaining suppliers in terms of helping them and monitoring their performance. Suppliers who used to cater to the lower levels of bills of materials are eliminated. They now comprise the second-tier suppliers and are managed by their customer who supplies the sub-system to the final assembler.

This facilitates a close relationship between the customer and supplier. This does allow more security to the suppliers but it makes their operations more visible and exposed to customer demand for continuous improvement in terms of cost, quality and delivery.

The main reason stems from the need to be cost competitive. The increased proportion of sourced components gives the opportunity to reduce costs in several ways. The responsibility of the cost for holding stock is transferred to the supplier. This statement might imply that the supplier now holds the inventory but this only occurs if the supplier's manufacturing system is inefficient and no assistance is given to rectify this. There is also a demand for the supplier to be cost-competitive and pass on any productivity savings to the customer. Since the proportion of components that are sourced externally is increasing, the supply chain is the largest area to improve. This is achieved by allowing the customer to examine the costing structure of the supplier and working together to reduce costs. The supplier is integrated into the production schedule in order to reduce the cost of stockholding.

Another aspect of cost saving is that of specialization. The supplier no longer produces to a given design but instead is given the required performance and cost parameters and a free rein to develop the product. The trend is for the supplier to build sub-assemblies which reduces research and development costs and in-house engineering staff levels can be reduced. Another benefit is that the supplier's technical expertise can be utilized.

Supplier development process

Nissan represents the present benchmark in the UK for customer–supplier relationships in the European automotive industry. It has only 160 suppliers compared to Rover, who has 350. Nissan has introduced into the UK its supplier quality assurance system which is similar to that it implements in Japan.

The supplier development process comprises three stages:

1 **The proactive stage:** this consists of agreeing to a standard in terms of price, quality and delivery and adhering to it. The emphasis is on proper

and thorough project management and planning. This stage is where a new supplier is being initiated or a new component/new model is introduced.

2 **The interactive stage**: this is the dynamic, in-production period. The competence and integrity of the supplier is monitored and efforts are made to help resolve any problems. Problems are regarded as an opportunity to improve and not for recrimination.
3 **The reactive stage**: the process of continuous improvement. *Kaizen* teams undertake improvement projects and the emphasis is on people involvement. These teams are not management driven, problem-solving task forces. In a similar vein, Nissan embarks on a programme of supplier development by giving direct assistance towards quality and productivity improvement. Targets are set and the benefits are shared by both parties. Sometimes, Nissan finances capital equipment required for substantial improvements on very favourable terms.

Nissan has been very successful with its suppliers through forging a close working relationship, to systematically plan, do and improve. Suppliers may not make high profits but they make profits which are stable and certain. Nissan's claim of being the benchmark in distinguishing its supply quality system from other manufacturers is the fact that good planning and project management is required.

Supplier performance assessment

An evaluation questionnaire is used by Nissan to address the following areas:

1 Management
2 Quality manual
3 Drawing/document and specification control
4 Control of purchased material
5 Control of process/finished parts
6 Calibration of gauge and test equipment
7 Control of non-conforming products
8 Machine control.

The auditors involved write their comments and score these aspects accordingly. Provision is made on the form for the supplier to reply and, where necessary, record what pertinent action is to be taken and time-scales for corrective actions to be implemented.

Kodak

Kodak's dedication to total quality management in the manufacture and supply of all its products and services extends from the marketplace, through manufacturing to its suppliers. Kodak recognizes that its quality objectives cannot be cost-effectively achieved without ensuring consistent quality of incoming materials and services. To achieve this, they have found it mutually beneficial to enter into a quality partnership with key suppliers to achieve continuous improvement in the quality of their supplies to Kodak. This forms the basis of the Kodak 'Quality First Supplier Programme'.

The 'QI' Programme is structured, coordinated by the purchasing department, in which key suppliers are invited to work closely with the relevant Kodak users and buyers in improving the quality and value for money of the materials and services they supply. The programme identifies and tackles individual areas of opportunity for improvement on a project priority basis. This is accomplished through small dynamic teams, involving the skills and resources of supplier personnel in conjunction with the Kodak purchasing and user departments. Working together, these team efforts are directed towards total quality improvement in all aspects of the business relationship.

Sharing of information, cooperation, trust and mutual understanding are fundamental to the programme and create a climate in which the supplier is encouraged to employ innovative methods of problem solving and quality control which will enhance the quality of the product/service supplied.

Supplier development

The characteristics of an ideal Kodak supplier are defined as follows:

- Products are 100 per cent correct and reliable
- Deliveries are always on time
- Quantities delivered are always correct
- Deliveries occur frequently to minimize stock carried by user
- The supplier provides appropriate response to urgent requirements
- If something goes wrong, there is total commitment to rectifying it as soon as possible
- Competitive product pricing
- Invoices and documentation are free from errors
- The supplier is totally open and honest about processes, costs and pricing methods
- Supplier works with Kodak to continuously improve performance.

Supplier development is conducted in a series of progressive stages and the supplier is accorded a particular status during each stage. These are defined as follows:

1 Potential
2 Accredited/approved
3 Level 1 (preferred)

4 Level 2
5 Level 3

Potential

This stage is where quotations and samples are provided for various specifications and trials are conducted.

Accredited/approved

Visits are made to suppliers and specifications are agreed and set. Limited business is conducted and the supplied products are subjected to inward inspection. After this stage, a formal system of awards is implemented. This comprises three different levels of achievement. To qualify for these awards, a supplier would be expected to participate in a quality management programme that entails regular buyer/client/supplier quality team meetings. These are as follows:

- Identification of opportunities for improvement
- Establishment of appropriate improvement projects and goals
- Measurement, monitoring and review of progress.

Level 2 (Preferred) – Quality First Preferred Supplier Certificate

This is the basic level award and demands that high levels of conformance to specifications and supply are consistently maintained and no significant failures occur over a defined period, normally 6 months or a year. Performance should be monitored by means of a checklist or a performance matrix. Separate criteria are used for materials and services.

Level 2 (Product Control) – Silver Award/Service Award

To qualify for awards at this level, a quality team (QT) should be appointed. In addition to the criteria for Level 1, confidence in a supplier's ability to consistently meet requirements is such that Kodak's incoming inspection and testing of goods is dispensed with. This state is referred to as 'product control' and responsibility for conformance rests solely with the supplier. Quality of design enhancements and other quality improvement should be realized by meeting the quality team's continuous improvement objectives. Nominations for a Silver Award must be accompanied by a performance matrix and additionally a 'self-audit checklist' completed by the QT unless the supplier has a recognized third-party accredited quality system (e.g. BS 5750).

Services providers will have proven their ability to provide a significantly high level of service over a sustained period and judged to provide a fault-free service. The supplier will have reached goals and objectives set and monitored by the quality team. The quality team shall also have completed the 'self-audit checklist', unless third party accredited.

Level 3 (Process Control) – Gold Award/Service Excellence Award

In addition to the criteria for Level 2, the supplier utilizes statistical quality monitoring and data analysis techniques to control all the key process parameters and has a minimum process capability (Cpk) of 1.3 for these parameters. This state is referred to as being in 'process control'. Nomination to this level also requires a satisfactory formal assessment of the supplier's quality systems by Kodak quality control specialists, unless third party accredited. In addition, a high level of proactiveness and dedication to continuous improvement should be evident (i.e. exceeding the criteria).

Service providers are expected to provide outstanding contributions in all areas of the business relationship, being proactive in proposing improvements to goods, services, methods and design. Again, all continuous improvement objectives and goals should be exceeded and there is the requirement to have a satisfactory assessment of the supplier's quality systems.

Supplier performance measurement

The key result areas that Kodak address when assessing suppliers are:

Quality of conformance to specifications
Quality of design
Quality of service
Statistical process control.

The tool used for measuring these dimensions is a performance matrix. This is the primary means of tracking progress in a Kodak Quality First Supplier programme. The matrix is used for:

1 Benchmarking current performance
2 Establishing the quality team's objectives and goals
3 Monitoring ongoing progress
4 Communicating progress within the team and to Kodak/vendor management
5 Summarizing the programme for award-nomination purposes.

Various performance dimensions are assessed on a scale of 1–10. These dimensions should satisfy certain criteria to ensure that they are fair.

Expectations

Suppliers can expect from Kodak:

- Early involvement in the design and establishment of its requirements
- Early involvement in the optimization of the production process
- Realistic and understandable specifications

- An accurate forecast of Kodak's needs and early sharing of pertinent information
- Timely payment
- Open, timely and accurate communications of any change in plans and requirements and any less than satisfactory condition in any phase of the customer/supplier interface
- Involvement in and support for joint customer/supplier improvement activities
- Recognition for meeting or exceeding Kodak quality objectives.

Kodak expects from its suppliers:

- Commitment to continuous improvement
- Conformance to contract agreements: on-time delivery, delivery to proper location, proper quantities, proper labelling, package, etc.
- High-quality, invariant products and services which meet Kodak's specifications and require no receiving inspection and extra cost
- Cooperative management
- Mutual sharing of the benefits achieved as a result of supplier/customer team improvements
- Open, timely and accurate communication of any process change and any less than satisfactory condition in any phase of the customer/supplier interface
- Supplier management that is committed to exploring 'state of the art' technologies in their field of endeavour.

The achievements gained by both Kodak and its suppliers are shown in Table 8.6.

8.5.3 Ten golden rules of supplier partnerships based on best practice

The various cases discussed cover specific areas which are thought to be very pertinent and critical for effective partnerships with suppliers. If one is to draw

Table 8.6

Suppliers	*Kodak*
Increased effectiveness	Optimum product design
Lower total cost	Increased effectiveness
Increased business opportunities	Strengthened long-term relationships with suppliers
Formal recognition by Kodak	Receipts conform to specifications Lower total cost

on 'best practice', then there are ten areas to be considered in the development of effective partnerships with suppliers:

1. Supplier quality statement, demonstrating visible commitment to work effectively with suppliers.
2. Integration of supplier development in company quality policy: to determine visible commitment towards building a long-term and lasting relationship with suppliers.
3. Defined supplier characteristics or requirements. The working relationship with suppliers has to be very explicit and based on expectations and outcomes.
4. Structured development strategy: how suppliers are going to be developed and integrated as partners.
5. Supplier performance measurement: the use of various indicators to monitor supplier overall performance against set targets and agreed parameters.
6. Extent of performance measurement: covering both hard and soft aspects and also short- and long-term issues. Measurement should also be seen as a trigger for continuous improvement and not a controlling mechanism for blame and reprimand.
7. Defined supplier expectations: partnerships are based on a win–win situation and customers need to be sensitive to the expectations of their suppliers.
8. Defined statement as customer/partner: similarly, suppliers need to explicitly indicate their visible commitment to the development of partnerships.
9. Structured supplier merit schemes: a need to recognize supplier efforts in meeting and exceeding quality standards and to encourage their innovation and creative potential.
10. Achievements: track and record achievements and publicize them internally and externally.

8.6 Benchmarking TQM in the financial services sector using self-assessment

8.6.1 Introduction

The financial services industry is in a period of major structural change and a number of factors demonstrate why this is so. Progressive deregulation of the building societies in the last 10 years [19–21]; the maturing of the home loans market [22]; change in the nature and structure of other traditional markets in personal savings, life assurance, and general insurance, [20, 22, 23]. Others have argued that global barriers to entry are eroding, adding to the competitive environment [24–26].

As a consequence of these events the major institutions are seeking to consolidate their positions, with a number of substantial mergers announced in

1995 and 1996 and several mutual building societies and insurance companies planning conversion to limited company status (Table 8.7). Research by Altunbus *et al.* [27] supports the notion of consolidation finding that reasonable scale economies are achievable in particular for bank–building society mergers. McKillop and Glass [28] also find evidence to support economies of scale and scope. Aitken [29] however, suggests that the main motivation for merger is largely defensive to secure market share.

Those organizations remaining with mutual status have come under intense pressure from members and outside speculators, resulting at times to suspension of some core trading activities, in the wake of windfall payoffs to those members of organizations converting or merging. Those remaining committed to mutuality are fighting back with major marketing campaigns and lower pricing of services. At the same time, recent research by Wilkinson *et al.* [30] finds evidence of customer dissatisfaction and bewilderment, and complaints to the banking and insurance ombudsmen are at rising levels.

These extreme environmental changes are in turn pressuring executives from the major organizations to re-examine their organization structures and business strategies with a need to seek structures, strategies, and systems which are more attuned with the new dynamic conditions. The pressures for better management of innovations, customer service, culture change, and process efficiencies are growing. The financial services industry has reached a watershed in its history.

8.6.2 The case for TQM in financial services

Modern management theory supports the notion that TQM is becoming established as the new management paradigm. Grant *et al.* [31] argue that the only generic and strategic alternative is the economic model of the firm [32(a)], of which they are critical for being essentially static and too narrowly focused on profit maximization and shareholder wealth. TQM, they argue, is more attuned to modern and dynamic business conditions, offering prescriptions on best

Table 8.7 *The UK banking sector*

	Total assets £m	*Mortgages* £m	*Branches*
Barclays	159363	14288	2067
Nat West	156893	11900	2416
Lloyds + TSB + C&G	124603	39460	2796
Abbey Nat. + N&P	103316	57314	999
Halifax + Leeds	92711	74816	1441

Source: published accounts

practices, an orientation to customers, and a balanced stakeholder perspective for measuring business performance.

Hill and Wilkinson [32(a)] state that satisfying customers is now the most significant performance measure guiding modern businesses. This supports the wealth of literature from the gurus of the quality movement who have long argued for this business focus. They conclude that the principles of Deming, Juran, Crosby and Feigenbaum are now widely accepted as management common sense.

Several studies have demonstrated the growth of the TQM principles among leading corporations in the USA [31–35]. Greene [36] concludes that TQM is now the management 'norm' among the top US corporations.

The growth of TQM methods in the West has undoubtedly been aided by the popularity of the self-assessment frameworks and award schemes such as the Malcolm Baldrige (in the USA from 1988) and the European Quality Award (from 1990). The European Foundation for Quality Management report over 500 member companies (of which notably in the context of this research several are major banks, building societies and insurance companies).

The growth of TQM seems not to have been seriously halted despite some very mixed results from research (Table 8.8). The studies by Witcher [37], Kearney [38] and The Economist Intelligence Unit [39] all conclude that the majority of companies studied remained committed, viewing progress as essentially a long-term process. Problems have tended to be attributed more to lack of management expertise rather than a failing of the basic philosophy [32(c), 40].

On balance there is a strong case to argue that financial services organizations should be looking seriously at TQM, yet the evidence suggests that they may be lagging behind other sectors. For example, in our initial survey of senior managers 60 per cent of respondents said TQM would be a vital part of their organization's strategy in the next 3 years. However, 75 per cent were concerned at a lack of expertise, and 70 per cent complained of top management's lack of commitment. One possible explanation for this gap may be as Dean and Bowen [41] describe, that the services sector has been slow to grasp the potential of TQM, deriving largely from a manufacturing sector background with its roots in production management and statistical process control.

8.6.3 Methodology

The leading group of organizations in financial services were identified following an extensive review of the literature, and exploratory research which involved:

- A survey of the major banks and building societies (thirty-five responses from fifty)

- Expert interviews with senior executives in the industry
- An analysis of corporate reporting over a three-year period.

From this research twenty organizations were subsequently selected for in-depth case study.

Each case study involved measurement of TQM following the EQA self-assessment model, which provides a framework for TQM and a methodology for evaluation. The advantage of using this approach was that it enabled contributions to be taken from a wide cross-section of individuals within each organization, involving people in discussion, debate and eventual evaluation. From these deliberations a consensus view would emerge. The drawback of self-assessment models is that they provide only simplistic factors as a guiding framework which critics argue are non-empirically based. However, this weakness has been diminished substantially in this study by inclusion of empirically based criteria within the framework incorporating the critical success factors from studies by Black [42], Saraph *et al.* [43], Bossink *et al.* [44], Porter and Partner [45] and Ramirez and Loney [46].

For the purposes of this section the results of the self-assessments are those arising from the enabler factors only, i.e. these include areas of leadership, policy and strategy, resources, people management, and processes and scores are out of a total 500 points available. In terms of the evaluation of business results different measures and criteria were used by the organizations involved, and hence to introduce greater objectivity (and to facilitate wider

Table 8.8 *TQM implementations: research findings on impact on performance*

Name/date	*Method*	*Sample*	*Findings*
Kearney (1992)	Survey	100 UK companies	80% initiatives fail to deliver expected benefits
London Business School (1992)	MBNQA assessment	42 UK companies	Low scores 100–400 out of 1000
Economist Intelligence Unit (1992)	Case studies	50 EU companies	Problems with implementation, many initiatives fail
Durham Business School (1992/1993)	Survey	235 companies North/ 650 Scotland	TQM still in its infancy/most organizations stay committed
Bradford University (1994)	Measured financial performance	29 known TQM companies	Better than average financial performance

comparison with industry performances) it was determined to focus purely on the impact of business performance using traditional financial ratio analysis. Measures of profitability, capital strength, market share, and cost containment were taken over a 3-year period. These performance measures were then compared with an industry index comprising the largest twenty building societies over the same period.

8.6.4 Findings/discussion

1 Where are financial services in relation to TQM?

Figure 8.7 illustrates the typical profile of the pathway to award status following the EQA model. It is based on the accounts of past award winners Xerox (1992), Milliken (1994), and Texas Instruments (1996). It also draws on research in the USA which has examined profiles of organizations over several years following Malcolm Baldrige [32(b)–34]. Typically, organizations following self-assessment procedures encounter few difficulties in the early stages and make reasonable progress in the first two years. First assessments for organizations new to TQM typically will score 100 points or less (on enabler criteria only). This can usually be improved upon by identifying and attacking weaknesses, and progressing to 200 points at 2 years. At this stage subsequent rounds of self-assessment become progressively more difficult as the organization starts to identify more deep-rooted areas for improvement and fixing these will take longer. It may take between 4 to 5 years to reach 300 points. Support from top management will be essential to keep the momentum going. Research from the USA suggests that at this point organizations

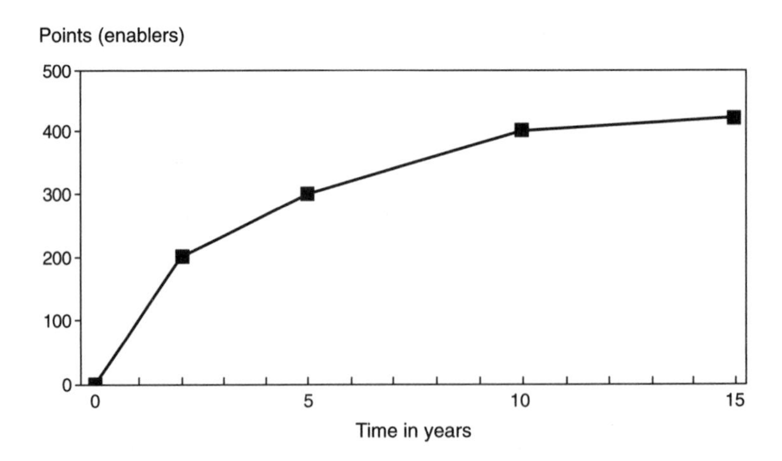

Figure 8.7 *Self-assessment: the path to award status*

encounter a barrier to further progress with as many as two-thirds failing to move on to higher points scores and potential award status. Reasons for this are attributed largely to a running out of steam; organizations become satisfied or simply stuck with their progress, or disillusionment sets in as progress appears more difficult (Figure 8.8).

Figure 8.9 illustrates the results from this research identifying the relative positions of the twenty organizations. The majority were assessed at below 200 points and most had less than 2 years' experience of implementation. A

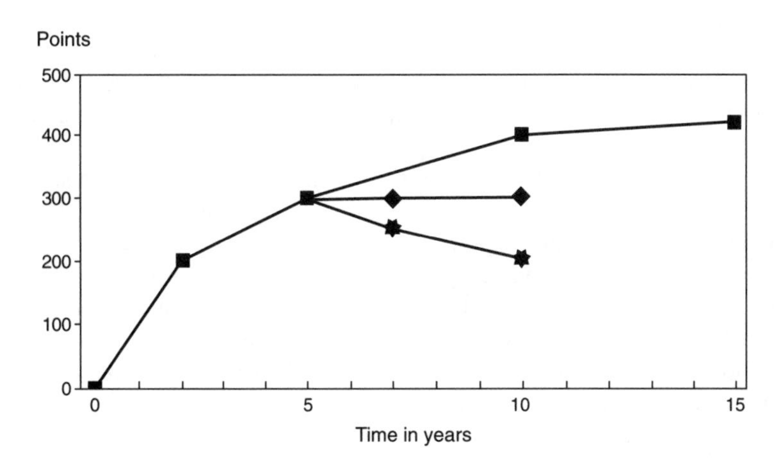

Figure 8.8 *Self-assessment: the 300-point barrier*

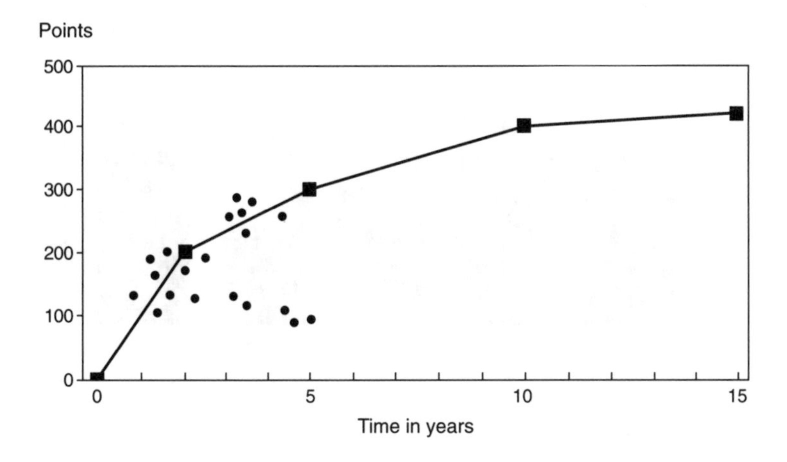

Figure 8.9 *Self-assessment: twenty cases from financial services*

small group of six organizations had progressed beyond this and were rated at between 200 to 300 points, with 2–5 years' experience. Of this leading group, three organizations were building societies (two of which have been the subject of recent takeover), two were specialist banks, one providing international services for retail and corporate customers, and one a major provider of cash-handling services. The remaining organization is a retail bank, also the subject of a recent takeover, and which achieved varying scores within its different business units, i.e. some SBUs were rated at around 300 points. However, this was not a reflection of the organization as a whole.

No organization in total emerges as a potential award winner overall, though a small number of SBUs within the ranks of the major retail banks are making progress towards this. For others their are some signs that activity may be levelling off or declining, and the extent of recent merger activity can only add a further layer of complexity to possible further progress.

Finally in this section the results of the leading group are compared against the best EQA submissions (top 10 per cent, 1994), by enabler factor. Figure 8.10 illustrates a significant performance gap across the range of factors. However, relatively high scores are achieved in relation to process factors, emphasizing a focus on this area in recent years with investment in information technology and business process re-engineering programmes. The results also show, however, that financial services are relatively weak in the areas of people management and in commitment of resources for quality improvement. Typically in relation to people issues the research finds that:

- TQM and HR policies were often incompatible
- There was discontentment in staff and senior management with practices

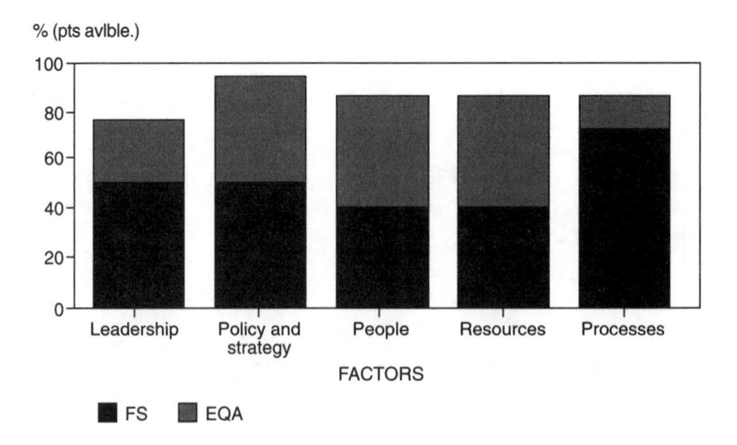

Figure 8.10 *The best by factor*

such as performance-related pay, personal appraisal, commission schemes, and top-down target setting.

These findings support research by Wilkinson *et al.* [30], which calls for a radical reappraisal of HR practices within the sector.

In relation to the commitment of resources the research finds:

- A lack of integration of TQM with the strategic planning process
- A lack of commitment and involvement of top management.

The impact on business results

The leading group were measured against a performance index benchmark of the largest twenty building societies. Figure 8.11 illustrates the results over a 3-year period in relation to change in market share, profitability, and cost containment. On average, the leading groups achieved significantly better than industry average across the range. In relation to increase in market share the leading groups were between 30 per cent and 50 per cent better than industry average over the period. In relation to improved profitability they achieved between 25 per cent and 28 per cent better and costs improvement by between 5 per cent and 8 per cent better. Two of the building societies from this group had been identified as major improvers across a range of measures by independent analysts UBS [47].

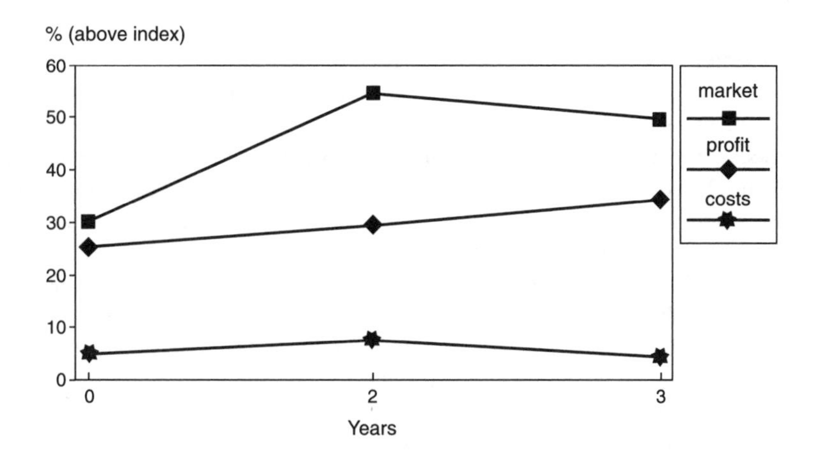

Figure 8.11 *Performance of leading group against industry average*

The lessons on best practice, critical success factors, and barriers

This study finds that the majority of the major institutions in the financial services sector are involved in some form of TQM-related activity. A few are taken with the underlying philosophy and are serious adopters, and some business units are near to progressing into the latter potential award-winning stages of self-assessment. Overall, however, the industry is in the early stages of adoption and has some way to go to match other sectors.

Several different approaches to implementation were evident with different levels of success (see Longbottom and Zairi [48] for a more detailed discussion of these). The main emphasis appears to be moving towards the use of self-assessment methods using either the European model or the Malcolm Baldrige. Several of the major institutions are now members of the EFQM, and many others have experience of completing at least one in-house self-assessment. The reasons for the popularity of such methods are not difficult to follow – the models are reasonably simple yet flexible and non-prescriptive in nature which permits tailoring of criteria to suit the organization. The process of assessment involves teams of managers and staff in identifying and facilitating the carrying out of improvement activities which need not, in the early stages at least, cause any major organizational changes or extra cost. Improvement activities need to be aligned with strategic objectives. The emphasis on measuring performance improvements switches from purely financial to a more balanced stakeholder perspective, taking account of customers, staff, the community, and process improvements, in addition to more traditional financial measures. This has prompted some to use balanced scorecard methods [49]. Such methods are popular and effective but need to be carefully integrated with the overall approach to quality, and care needs to be taken to avoid the temptation of top management reverting to imposing targets in difficult times, which may undermine the whole process.

Research has shown that many TQM initiatives fail in the long run. The process of improvement is a long and difficult journey. In using self-assessment methods there appears to be a barrier to progress at around 300 points, 3–5 years in. This research finds that organizations may be tempted to launch self-assessment programmes with inadequate preparation. A more considered and diagnostic approach is recommended, requiring the organization to audit its strengths, weaknesses, attitudes, skills and competencies before launching. In this way the organization is more likely to be able to select a strategy which is more closely aligned with its purpose and core competencies. It may also identify at an early stage shortfalls in skills and competencies which will need to be addressed with training, education and recruitment.

Table 8.9 summarizes the principal factors and best practices which practitioners considered to be most crucial for successful implementation. However, it was evident from the cases examined that in many instances practice did not follow these prescriptions. Table 8.9 shows a comparison of best practice arising from the review of literature, from practitioners own views and experiences and from observation within case studies. For example, leadership and involvement of top management has been stated to be a critical element from the literature, and participants in this study also regard this as a crucial element for success. The reality is, however, that in many instances top management are not involved.

8.6.5 Conclusion and recommendations

Arising from the research a new model for the implementation of TQM is recommended (Figure 8.12). The model promotes the use of self-assessment and balanced scorecards and four essential stages of diagnosis, implementation, measurement and review are recommended. Within each of the stages an approach and methodology have been devised arising from the principal findings from the research on the critical factors for success, best practices, and barriers to implementation. It is believed that by more careful diagnosis of organization strengths and weaknesses at the outset, and by careful selection of appropriate strategy, organizations stand a greater chance of successful implementation in the long run. At present too many initiatives are faltering in mid-term, as a result of strategic mismatches with core competencies.

8.7 Best practice in the car aftersales service: an empirical study of Ford, Toyota, Nissan and Fiat in Germany*

Olajide Omotuyi Ehinlanwo, Iomega Europe GmbH, Germany and Professor Mohamed Zairi, University of Bradford Management Centre

8.7.1 Introduction

The value system in the German car industry can be broken down into four major players; component suppliers, car producers, dealers or the point of sale, and buyers. These players are all affected by any turbulence in the car industry. As a result they all contribute, to various extents, to the attempts of structural change taking place in the industry. As Müller and Reuss have argued [1], perhaps what is most noticeable of the attempts at readjustments within the value system is the insecurity on the part of all players as to what the future direction of the industry will be. Some of the major trends or key factors affecting each of the key players are:

*Reproduced with kind permission from MCB University Press Ltd.

Table 8.9 *Summary of best practices/critical success factors from case studies*

Factor	*Best practice*
Leadership	• CEO commitment, lead TQM, involvement not delegation • Understanding and expertise in TQM critical factors, principles, techniques • Develop quality initiatives to suit the organization – do not rely on packages • View TQM as a strategic tool • Emphasize long-term investment in education, training, research, innovation • Understand key business processes/hands-on style • Take long-term and balanced stakeholder view of performance • Favour coaching style, pragmatic
Policy/strategy	• TQM part of corporate planning process • Integrate TQM into business and business plans • Focus on deployment – reduce communication lines, use cross-functional teams, team emphasis not individual appraisal • Emphasis on corporate values • Use self-assessment • Use balanced performance measures
People	• Align HR policies and practices with TQM • Increase emphasis on teams • Commitment to long-term education and training • Emphasis on roles and competency development • Feedback and open communication with staff – emphasis on action on key issues • Increasing emphasis on team-based rewards, competency-based pay, gain sharing • Decreasing emphasis on personal appraisal, performance-related and commission-based pay • Focus on changing activities rather than more abstract cultural issues
Resources	• Integrate TQM into business • Focus on deployment • Employ experts • Reduce lines of communication • Use project and cross-functional teams • Develop flexible organization structures • Focus on roles and competencies rather than individuals and functions • Develop external links – for expertise, for benchmarking, for supplier relations

Table 8.9 *continued*

Factor	*Best practice*
Processes	• Identify and focus on key business processes • Develop critical performance measures – few but relevant • Develop process organization culture • Hands-on style from top • High use of principles – prevention, empowerment, continuous improvement, SPC and measurement, teamwork • High use of techniques – BPR, benchmarking, QFD, flowcharting quality systems
Performance	• Moving to balanced measures/stakeholders • Moving to self-assessment frameworks, e.g. EQA/MBNQA

- **Car suppliers**
  Polarized competition
  Internationalization and globalization tendency (for example, the global sourcing producer strategies and global competition, etc.).
  Change from just component suppliers to system partners and suppliers integrated into the system.
  Integration into the producer system-based total quality management process.
- **Car producers**
  Concentration of corporate strategies on product, process innovation and cooperation.
  Emphasis on customer loyalty and development of repeat buyers at point of sale.
  The choice of place to manufacture and its inherent implications and cost savings strategies has become a key focus of corporate strategies.
  Global thinking and globalization.
  Reduction of overcapacity.
  Concentration on return on investment (ROI) improvements.
  Continuously expanding model width and depth.
  Shorter model lifecycles.
  Increasing use of merchandising tactics.
- **Car dealers**
  Overcapacity of dealers in the European market especially in Germany.
  Concentration tendency among dealers (takeovers, mergers and fusion).
  Reductions in profit levels and return on sales, etc.
  Increasing capital requirements.
  Undercapacity and underutilization of facilities.
  Loss of market share in used car, and parts and accessory (aftersales) business.

Table 8.10 *A comparison of key findings: survey/case study/literature*

Critical factors	*Questionnaire (respondents' rating)*	*Case studies (evidence shows)*	*Research/literature (evidence/best practices)*
Leadership	Important	Not effective	Critical for success/change of 'mind set' needed in 'West'/some 'green shoots' in USA
Mission & strategy	Important	Not integrated/not high-level activity	Must be high level and strategic and integrated into business
Organization structure	Uncertain	Favour functional structures/ emphasis on culture change	Focus on critical success factors/favour cross-functional teams/emphasis on activities aligned to strategic objectives
Human resources	TQM/HR must be consistent	TQM/HR not consistent/ in conflict	Need for greater integration of TQM/HR policies and practices in 'West'/inhibiting progress
Processes	Important	Processes narrowly defined and limited/emphasis on short-term cost reduction and incremental change/ restrained by functional boundaries	Broader definition of processes (e.g. HR, marketing, strategy, innovation, etc.)/emphasis on principles of empowerment, prevention, continuous improvement, best practice
Performance measurement	Very important	Focus on financial measures/short term emphasis	Emphasis on balanced measures/long term

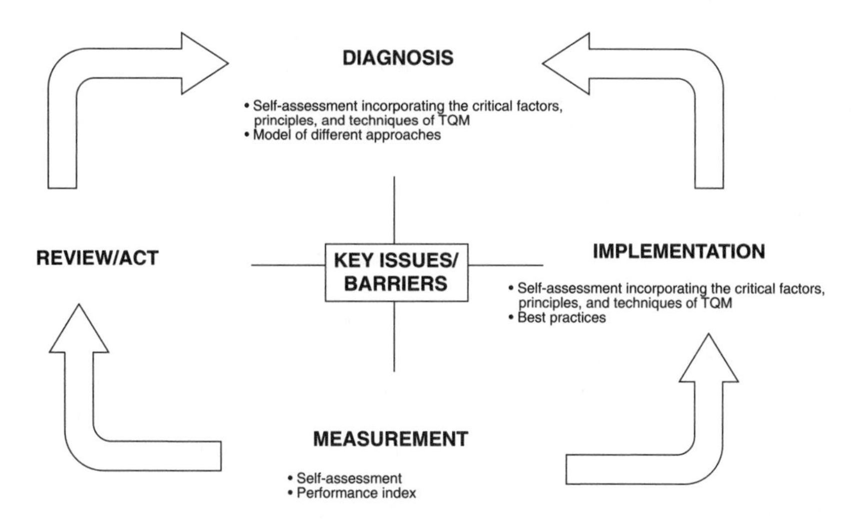

Figure 8.12 *A new model for TQM implementation*

New company forms and competitive forms, e.g. Mega dealers – 123, Pit Stop, etc.
Pressure from producers to increase product intake levels.

- **Car buyers**
  Stagnating/cyclic demand in European market.
  Stronger price/quality awareness.
  Increase in service requirements and importance of the service part in the buying process.
  Increasing difficulty in telling the difference between available models.
  A wish for totally new car concepts (expectations of better security, environmental considerations, etc.).

Underlining all these trends and the competitive pressure is the realization in the car industry that the present downward trend in volumes can only be stemmed by an increase in customer loyalty and the development of stronger brand/corporate awareness among customers. The stagnating/cyclic demand, stronger price/quality awareness, increased emphasis on service and general economic pressures which have reduced buying power have necessitated the need to develop strategies that are geared towards achieving greater brand loyalty among customers. There is, more than ever, a need to ensure that present customers are not only satisfied but also continuously delighted.

The aftersales sector, long after the customer has taken delivery of the car, represents a means of continuous contact between the car producers and the customers via the dealers. As a result of the unique vertical marketing system practised in the car industry and the fact that the balance of power lies with the car producers, they have been forced to and must assume leadership in an aftersales process geared towards retaining the end customer through total satisfaction. Whatever aftersales concept they sanction, it must optimize both the delivery of the product and the service aspect of the process (see Figure 8.13).

8.7.2 Aftersales definition and scope

Without laying any claims to being all-embracing, aftersales service for the purpose of this section would be defined as 'all activities geared towards maintaining the quality and reliability of the car carried out after the customer has taken delivery with the goal of ensuring customer satisfaction'. These activities can be sub-divided into those that take place at the producer level and the dealer/point of sale level. This section specifically examines the activities at the producer level. While these activities are channelled towards a complex range of outlets, for practical purposes the examination has been specifically limited to include only dealers contractually bound to a car producer.

In the boom years until 1991, aftersales on the producer level consisted of three major departments; the parts (*teile*), accessories (*zubehör*) and service, mostly technical (*kundendienst*). This separation was seen as necessary partly because of the difference in profit margins to be made on parts and accessories but also because of the strategies of the importers in which accessory additions for the German market were made to imported cars on their arrival. These accessory additions (sometimes known as Rapid specs or *Erst-ausrustung* in the industry jargon) were booked as parts or accessories sold for accounting purposes. However, for the purpose of this section, those component additions made by the German daughter companies of the importers, where possible, have been excluded from the analysis. These additions are only intended to bring the car to the level expected in the market. Aftersales would, therefore, be seen as having a product (*teile*, *zubehör*, etc.) plus a service component

Figure 8.13 *Car repeat buying in relation to customer satisfaction. Source: adapted from Moritz Spilker [51]*

(additions or changes or advice carried out after delivery of the car to a customer). **Product** is defined as all components that can be fitted to the car to enable it to meet the requirements of the owner and **service** that human or mechanical component/interaction required to install or make aware of the possibility of installing a product on the car. These include advice, information, and all interaction with the car owner/user.

Products can be sub-divided into four main groups:

1. **Parts**: These are the components of a car that, due to long- or short-term wear and tear, have to be replaced to ensure that the car fulfils its primary function, i.e. serves as a means of transport.
2. **Accessories**: These are the extras that can be bought and installed in a car that in no way affect its performance or its primary function.
3. **Tyres**: This category includes both summer and winter tyres as well as normal and wide tyres.
4. **Autochemicals**: These are the chemicals that, apart from serving to help ensure the fulfilment of the car's primary function, are also used to clean and maintain its aesthetic qualities. This classification includes car paint and car paint accessories, motor oil, special car cleaning and car wax materials, etc. [52]

8.7.3 Conceptual framework

Problem definition

The process of change is very difficult and organizations tend to accept much more easily changes that have been tried and tested in their industry and found to work or contribute to success. The structural upheavals in the car industry and the problems both producers and their dealers are facing in improving the customer loyalty rates necessitates a search for both new methods and the revival and improvement of existing policies and processes. The use of total quality management organization principles is now a fact of car management. Benchmarking is one of the basic pillars of total quality management and it is the aim of this study to use benchmarking to examine the aftersales practices in the car industry and extrapolate best practices that could lead to overall improvement of the aftersales delivery as it currently exists.

The present trend in the car industry is to push for improvements at the dealer/point of sale level, and while these improvements are necessary so also is the realization that the aftersales policies and processes of the producer are basic to achieving these improvements. According to Müller [53], car producers must realize that 'achieving dealer satisfaction is both the basis of and a requirement for true end customer satisfaction'. Dealer satisfaction can only be achieved when the aftersales policies of the car producer take into account factors that are critical for continued dealer success. The critical policies as far as the dealers are concerned represent the marketing policies of the producers. The environmental considerations should be included in that they represent one of the most pressing challenges facing all car producers.

Critical policies

For every strategy, process or policy, there exists certain critical aspects whose fulfilment are critical to the success or otherwise of the process strategy or policy (Figure 8.14). Aftersales service takes place through the vertical marketing system also used for new car sales which is made up of theoretically independent partners. However, the system as practised in Germany can be further broken down into both the single- and double-phased system [54]. A single-phased distribution system exists when the producer, without the services of any other go-between supplies the dealers or point of sale. Practitioners of the single-phased system include BMW, Honda, Porsche, etc. In the double-phased distribution system the producer supplies a number of dealers directly and these in turn have one or more dealers' point of sale that they supply. Practitioners of the double-phased distribution system include Toyota, Fiat/Alfa/Lancia, Subaru, Volkswagen AG, Ford, etc.

Core activities in this marketing system revolve around the policies, processes and strategies employed by producers in ensuring that their representatives or outlets are adequately prepared and able to satisfy the customer. Based on a review of existing literature and interviews with industry experts and dealers, a number of producer policies were identified as being critical or crucial to the aftersales delivery. They are:

- **Product policies**: Dealers have on many occasions expressed concern at the product policies as practised in the car industry. Based on the GVO

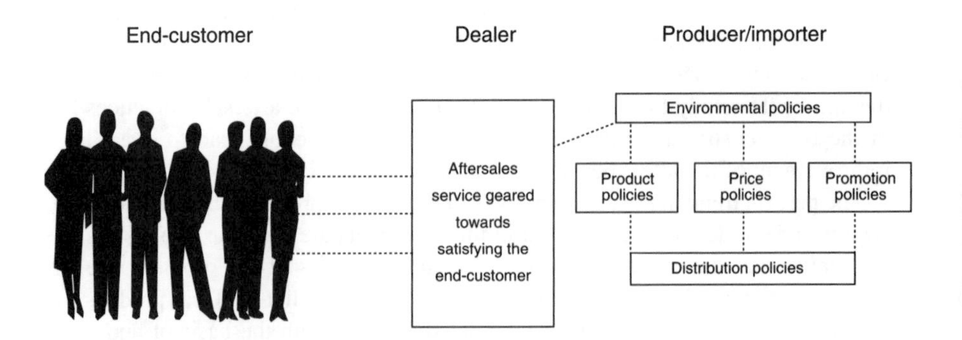

Figure 8.14 *Aftersales critical policies and policy links*

(Gruppenfreistellung verordnung, the European law regulating the vertical marketing system), car producers had the right to bind dealers to sell only their original parts or products. The dealer was therefore bound to buy from a car producer even when the prices were not favourable. In addition, after-sales service is made up of a product aspect, and the conditions under which the dealer gets this product certainly plays a very important role in his success or otherwise.

- **Price policies**: These policies affect the bottom line and as such are very important for both the dealer and the producer. One of the major problems facing the point of sale is the dealer image of being too expensive in comparison to other aftersales outlets. A key thrust of all producer policies therefore is an attempt to introduce transparent price policies at the point of sale. In addition, the end-price which the dealer offers aftersales service is, to a large extent, affected by the price conditions the dealer is offered by the producer.
- **Promotion policies**: In order that the target audience becomes aware of the aftersales possibilities offered by both the producer and dealers, a well-designed promotion policy is required. Even the best policies, if not communicated to the target audience, can be doomed to failure. Traditionally, in the car marketing chain, producers played a dominant role in the conception of promotional measures and the recent 'corporate identity' measures of most producers do demonstrate the importance of promotional measures to aftersales. The promotion policies of the producer are, therefore, critical to the success or otherwise of both that producer's aftersales policies and that of the dealers.
- **Distribution policies:** The availability of the aftersales products affect the aftersales delivery. The perceived quality of service is directly influenced by how long the customer must wait to have the desired product fitted to his or her car. For the dealer/point of sale, profitability is affected by the number of products that he has to stock in his stores. The lower these are, the less fixed costs he has to bear and the more profitable and liquid he is. The distribution policies and their efficiency, therefore, play a critical role in the aftersales plan.
- **Service policies:** The product car is becoming universal in offering types of technology. Service has become a way in which policies and offerings can be differentiated from those of competitors. Granted that all other critical policies function effectively, the service policies represent the most important differentiation technique available to both the producer and his dealers. With consumers becoming more homogenized in their requirements, it is of paramount importance, therefore, that service policies are geared to offer that little extra requirement to delight the customer. Under service, for example, would be included the training and consultancy activities of the producers geared towards helping their point of sales/dealers compete more

effectively and satisfy the end-customer. The competition on the point of sale level and the continuous technical evolution of the product car has made it critical for producers to offer their dealers a whole range of support and training measures, to help them secure the additional support required to both satisfy and bind the customer.

- **Environmental policies:** Consumer trends demonstrate the 'coming of age' of environmental considerations. In Germany, and indeed the whole of Europe, industry has been forced to incorporate environmental awareness and considerations into their policies. The car producers and their dealers can no longer ignore the importance of the environment in their future policies if they do not want to alienate their customers. Numerous surveys show that buyer behaviour in Europe and indeed Germany is influenced by environmental considerations and organizational policies.

Supporting these identified critical policies is the results of a study carried out by Professor Meinig of the University of Bamberg [55]. In this study, 1120 dealers contractually attached to twenty-eight producers were asked to rank aspects critical to the success of their operations and the achievement of customer satisfaction. The following aspects of aftersales were among the factors identified by dealers as critical to their success:

1 Aftersales price policies
2 Aftersales distribution policies
3 An adequate market size (i.e. the geographical area for which the dealer is responsible)
4 Aftersales service policies
5 Aftersales promotion.

8.7.4 Data collection and analysis

A representative sample of the car industry made up of practitioners of both the single- and double-phased system should have been prepared. However, as a result of time and resource constraints, this study had to be designed as a 'pre-test' aimed at identifying both areas for closer consideration and further study. Seven leading companies were initially identified: Adam Opel, Renault, Volkswagen, Fiat, Ford, Nissan and Toyota. Eventually only four companies took part in the study. These included: Ford, Nissan, Toyota and Fiat.

The questionnaire was made up of both open and closed questions. It was also necessary to design most of the questions to be of a qualitative nature as there was a general unwillingness among the subjects to release quantitative information. Furthermore, since the information required touched mostly on the drive for competitive advantage, some subjects that participated retained the option to refuse to answer any question which they perceived as too sensitive. The questions addressed both the enablers, i.e. the policies as they are practised and the results/effectiveness or otherwise of these enablers from the perception of the subjects.

8.7.5 Fiat AG Deutschland

1 Enablers and policies

General company information

As at 1994, the Fiat Germany group, i.e. including Alfa-Romeo and Lancia, occupied the number-two position in the 'car-producer–importers' list. It had a market share of 3.62 per cent (based on a calculation date of 1 July 1993). In a paper presented at the first 'Automobilwirtschaftliche' symposium in Bamberg (June 1993), Dietmar Fütterer, General Director Fiat Automobil AG summarized Fiat's strategy for the 1990s under four critical thrusts [56]:

- Improvement of the product development
- Improvement of the production process
- Improvement of the marketing process
- Improvement of the dealer network.

These laudable goals represent Fiat's approach to coping with the structural upheavals facing the industry. Of significance for the aftersales service are the last two measures or goals which are perhaps also a natural follow-up of the introduction of total quality management or 'qualita totale' at Fiat. As a result of a recent reorganization, Fiat assumed responsibility for the aftersales activities of Alfa-Romeo and Lancia, operating a two-phased marketing system. The marketing network is made up of 608 category A Fiat dealers, 300 category A Lancia dealers, 238 category A Alfa-Romeo dealers, and approximately 650 category B dealers (Figure 8.15).

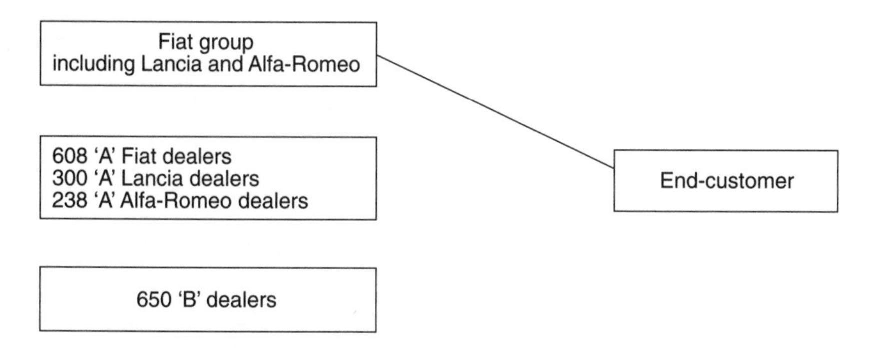

Figure 8.15 *Fiat's aftersales outlets*

Responsibility for aftersales service is borne by the department service resort reporting directly to an executive director service. The service resort is further sub-divided into marketing, new car sales, customer service and parts and accessories (Figure 8.16). The customer service and parts and accessories department is directly responsible for achieving Fiat's target of 'total customer satisfaction' with a staff strength of 147. This is broken down into fifty-nine internal workers at Heilbronn and eighty-eight regional workers. Of the eighty-eight regional workers, five are regional heads, ten troubleshooters, five secretaries and sixty-eight field representatives or dealer consultants. To achieve its target of total customer satisfaction, Fiat has introduced four new key measures in addition to existing policies in the examined critical areas of price, promotion, product, distribution and the environment. These are:

1 A three-year dealer development plan (1993–96)
2 The Patto Chiaro programme
3 The Front Line programme
4 Encouragement of dealer ISO 9000 certification.

Its major customer target groups are all Fiat/Lancia/Alfa owners with increased concentration on the re-acquisition of customers with cars older than four years whose loyalty rates are, as with other car producers, presently declining.

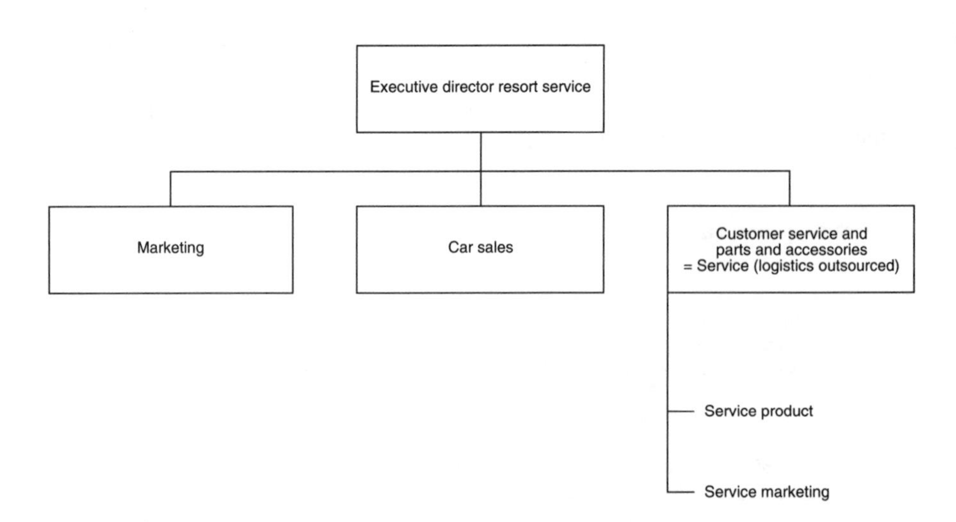

Figure 8.16 *Fiat's organization chart*

Product policies

Fiat stocks approximately 80 000 products which are sub-divided into 700 product families in its central store. It does not, however, stock all products that could be fitted to its cars for reasons of economy. According to Fiat senior managers, certain products have very low demand levels and using Pareto's 80/20 law, these products are unprofitable to stock in the central store. As a result of Fiat's close contact with original product manufacturers like Marrielli who also belong to the Fiat group, a number of its aftersales products are contracted out. However, up to 95 per cent of the aftersales products carry the Fiat brand name. Resulting from the variations in standard accessory requirements between European countries, approximately 95 per cent of Fiat Germany's accessory requirements are sourced locally in Germany. These products are then initially checked in Italy to ensure that they meet the laid-down Fiat quality standards.

A specific estimation of the percentage of Fiat's aftersales products that can be categorized as fast movers could not be given. However, approximately 70 per cent of the fast movers are manufactured by Fiat daughter companies and 30 per cent by Fiat. All Fiat products carry a warranty which can be sub-divided into a twelve-month warranty for technical defects on new cars sold, a three-year warranty for paint defects, and an eight-year warranty against rust. Furthermore, all aftersales products come with a twelve-month warranty except batteries, which have a two-year warranty.

As from April 1996, it is planned to offer customers the possibility of extending their warranty period for an extra fee. Fiat maintains that its warranty covers the cost of fitting. However, some of its dealers disagree. According to Fiat sources, the disagreement is as a result of dealers not excluding the product discount and bonus offered from their calculation of prices for warranty jobs. Fiat's products are, to a large extent, not compatible with other car producers' products when compatibility between the Alfa-Romeo, Lancia and Fiat is excluded. However, some of the Porsche group's products match Fiat's and it cannot be ruled out that some dealers, in attempting to satisfy the customer, use these products.

Distribution policies

Logistics at Fiat is outsourced to TNT Deutschland. TNT assumes responsibility for ensuring that the aftersales products arrive at the different points of sale when they are required. Normal supplies usually come from the central store in Heilbronn. In addition, Fiat maintains fourteen mini-stores in Hamburg, Osnabruck, Bielefeld, Düsseldorf, Niewied, Kassel, Frankfurt, Munich, Dresden, Freiburg, Ravensburg Berlin, Passau and Heilbronn which stock about 20 000 products for express deliveries. Only three of these stores, Hamburg, Frankfurt and Düsseldorf, are owned by Fiat.

'A' category dealers are integrated in a computer-based communication network called the DCS which is installed by Fiat free of charge. The dealers are able to place orders through this system. However, of the 'B' category dealers, only approximately 150 have computer systems and they have to pay if they wish to be connected to the DCS system. The majority of the 'B' category dealers order mostly from the mini-stores by telefax. The normal deliveries are by truck and take place weekly. About 40 per cent of deliveries are normal stock orders and 60 per cent of deliveries express. Both normal deliveries and express delivery costs are borne by Fiat. Express deliveries take between a minimum of eighteen hours and a maximum of twenty-four. The possibility of a mistake occurring in the delivery system was estimated at 0.01 per cent. The ordering system is a mixture of both computerized and manual procedures. Fiat does not have a compulsory minimum stock level for its dealers but recommends that they maintain approximately 400 fast-moving products in their stores.

Promotion policies

Eighty per cent of the support information is manual, made up of catalogues, product information leaflets, etc. Approximately 20 per cent of the support information contains required product fitting times. This information assists the dealer in estimating the times required for aftersales fittings. With this information it is then possible to better estimate the required times and give the customer a collection appointment. The practice is to generally inform dealers immediately of any changes in product or technical information. However, where changes do not occur, aftersales information is routinely updated twice monthly.

A complete recommended aftersales presentation concept does not exist. However, in the case of accessories, Fiat does have a presentation concept which it recommends to its dealers. No specialized aftersales advertisements are made but in reaching its target aftersales audience, Fiat employs public relation activities, regional and dealer- or area-based advertisements. The costs for these advertisements are split into 50 per cent paid by the dealer and 50 per cent paid by Fiat. Under the public relations activities which are closely tied to special service measures, the dealers are offered service mailings free of charge. Included in the mailings is a short letter, a service folder and a return envelope/card for the customer's answer.

In addition, under its 'loyalitätsmarketing' programme, Fiat has introduced a number of promotional measures aimed at both retaining its existing customers and re-acquiring the lost ones. Fiat designs for each dealer an individual brochure containing service offerings, competence statements and price information about the dealer. The first page is a photograph of the garage chief mechanic, followed by a full A4 page with calculated prices designed to prove that the image of the dealers as being more expensive than other aftersales outlets is not the case. Included in the brochure is also information on the additional advantages of using the services of the dealer in question with his address, opening hours and telephone numbers. This promotional measure closely tied to the so-called *Patto Chiaro* (we keep our word) measures costs the dealer DM395 per 2000 pieces. In addition, advice is offered on how to distribute the brochures to ensure that not only present customers receive this information but also those younger customers that tend to use other aftersales outlets as a result of cost considerations.

Service and related policies

Central to Fiat's service concept is the *Patto Chiaro* and the MOS-minimum operating standards or front-line programme. Under these programmes, dealers are given a list of helpful tips, hints and standards that are regularly measured and which is linked to a customer satisfaction index. The result is used to calculate an extra bonus given to participating dealers. All 'A' category dealers are free to participate. The MOS measures the dealer process when the car is taken in for repairs, the repair process, the dealer quality assurance process, the process by which the car is given back to the customer, including explanation of the billing and the warranty process.

In addition, the customer satisfaction index gives the dealer a feedback twice yearly on the satisfaction levels of his customers. Where there is a negative customer satisfaction index result, a diagnosis brochure made available from Fiat helps the dealer to determine possible reasons for the problem and to suggest solutions.

In addition, dealers and their personnel are offered training seminars both at Fiat and in specialized training schools. Topics covered include technical developments as well as commercial and customer contact methods. Approximately 90 per cent of the costs of this training is born by Fiat. The attendance rate among dealers for the compulsory courses is 95 per cent and for the non-compulsory courses approximately 70 per cent. Apart from the sixty-five sales representatives, two troubleshooters per region are also placed at the disposal of the dealers for prevention of possible problems at the point of sale. A telephone hot line is available within working hours, i.e. eight hours daily to provide answers and solutions to dealer problems.

Dealers also have a free option to take part in an ISO 9002 certification process jointly coordinated by Fiat and DEKRA, a quality certification organization, and based on ZDK (Zentralverband Deutsches Kraftfahrzeuggewerbe, the car association) service concepts. In addition to this certification process Fiat offers its dealers the opportunity to take part in a business management scheme in which key ratios and business results of dealers are compared with a view to discovering weaknesses and improvement opportunities. At the moment, about 400 dealers are taking part in the programme.

Price policies

Fiat does not offer the end-customer special aftersales financing schemes. However, dealers are given financial support in the form of credit on aftersales products ordered. In addition, for fast-movers, they are offered special credit opportunities with fixed repayment agreements. A discount or rebate system is also practised. The discount offered is dependent on the form of product delivery. On average, this discount is about 35 per cent of the selling price. Special bonus plans in place include the MOS bonus scheme and a bonus system based on the store turnover. All aftersales products are priced singly. The recommended hourly payment rate is DM91 on average and varies according to region and aftersales repair jobs, e.g. it would cost a lot more to carry out bodywork repairs. The average cost for aftersales service varies according to the model in question but, on average, costs around DM360 for a check-up on the smaller models and DM800 for the higher-class models. For regularly replaceable parts it costs, on average, DM80 and the end-customer pays about DM831 on average for accessories (all prices represent market prices).

Environmental policies

Approximately 70 per cent of Fiat's products are recyclable. However, Fiat does not at the moment have a special environmental concept for the aftersales sector alone. The existing concept is general and covers all its activities. There are plans to continue to introduce measures which are environmentally friendly and it is at the moment a member and party to the Eureka working agreement. Special arrangements exist for recycling batteries, windscreen and glass products as well as a catalyser programme in which end-users are encouraged to have their cars fitted with catalysers. Engine blocks are also collected and repaired or recycled at Fiat.

2 Policy effectiveness

Financial results

The complex accounting systems used in the car industry which separates aftersales products for accounting purposes made it difficult to secure total quantitative results for the aftersales sector. However, approximately 14 per cent or DM700 million of the last financial year's turnover came from the aftersales sector. Parts represented approximately DM570 million and accessories about DM130 million (market prices). Of these, the fast-movers represented approximately 23.3 per cent of the aftersales turnover. Accessories make up about 3 per cent of the total turnover with 80 per cent coming from the 'initial fitting' when the car is sold (*Erst-ausrustung*). The profit margin on parts is about 17 per cent.

Ranking of policies

Asked to rank the effectiveness of its policies, Fiat graded its product policies as very effective, its promotion, distribution and service policies as effective, its price policies as partly effective and its environmental policies ineffective (see Figure 8.17). These rankings were based on the justification that Fiat's products, as a result of the technological advances achieved in the production process, led to a technical advantage over competitors. In addition, dealer satisfaction with promotion policies has been achieved perhaps largely as a result of Fiat offering a range of free promotional support tailored specially for each dealer. Fiat considers its prices as a little above the average and the pricing policies being rated partly effective also reflects the trade-off between offering the dealer cheaper products and a wide range of service measures. Environmental policies were rated ineffective because of the non-existence of a special aftersales environmental policy.

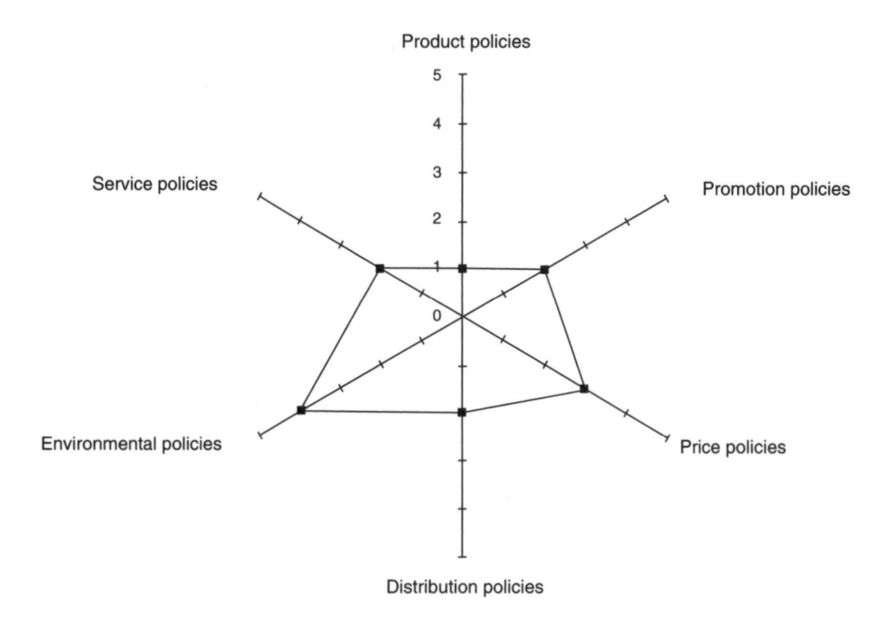

Figure 8.17 *Fiat's ranking of the effectiveness of its aftersales policies. Key to scaling: here and in following figures 1 = very effective (60–100%), 2 = effective (60–79%), 3 = partly effective (40–59%), 4 = ineffective (20–39%), 5 = very ineffective (0–19%)*

8.7.6 Nissan Deutschland GmbH

1 Enablers and policies

General company information

The Nissan company is Japan's second largest carmaker and one of the largest in the world. In Germany, it is the fourth largest importer with a market share of approximately 3 per cent. Nissan is one of the very few companies with an aftersales director. This might perhaps serve as an indication of the importance attached to the aftersales service at Nissan. It has recently launched a pan-European retail identity called 'Network 2000' – a reflection of the attempts in the industry to build a stronger corporate identity [57]. This programme includes, among others, a stylish exterior and interior design for dealer premises and facilities and, according to Nissan, sources an attempt to provide coherence between the quality of Nissan products and the environment in which they are sold and serviced.

The aftersales service is organized under an aftersales director with responsibility for the conception of aftersales strategies for Germany (Figure 8.18). This department is further sub-divided into seven units: sales support, marketing, regions and zones, training, product information, workshop and the environment. Nissan has 920 aftersales outlets in Germany, and fifty of these specialize in Nissan trucks and 870 are responsible for cars.

For Nissan's aftersales objectives, two goals or objectives were ranked as paramount;

1 To improve customer satisfaction through qualified operation, and
2 To improve profitability in dealers and the company.

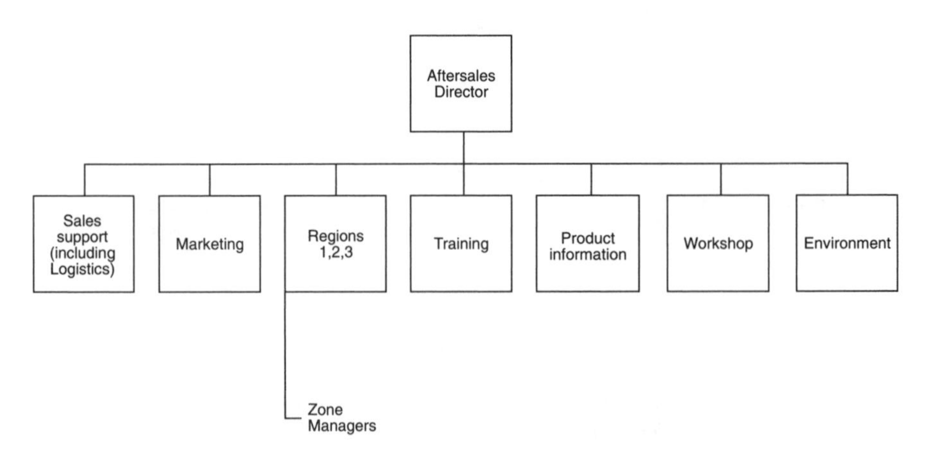

Figure 8.18 *Organization chart of Nissan's aftersales departments*

According to Nissan officials, target customer groups include all Nissan drivers and dealers. Twenty-five per cent of its aftersales staff strength work outdoors and 75 per cent work and coordinate directly from its German headquarters in Neuss.

Product policies

Amsterdam is the home of the vehicle and parts distribution operations of Nissan in Europe. The parts centre houses approximately 140 000 parts and over 480 000 inventory items. These parts are classified into five product families; the SSPI or competing parts to which 4000 single products belong; the Eye-Catching group (7000 products); the HUK or Body group (2000 products); the captive parts with the largest number of products (121 000); and the accessories (6000 products).

Nissan Germany stocks 33 000 single aftersales products in its German warehouse or Master Depot at Neuss. Nissan does not, however, stock all aftersales products that its models may need because of economic reasons. Products not stocked are those where demand falls below a 3 per cent monthly average. Ten per cent of its aftersales products are contracted out, i.e. not manufactured in wholly owned production facilities. However, all Nissan's aftersales products carry the company's brand mark/name.

Two per cent of the aftersales products are classified as fast-movers, i.e. products that are often in demand. These fast-movers are up to 95 per cent manufactured by Nissan in wholly owned production facilities. All Nissan's products carry a warranty. Although the warranty for the first three years covers the car, it also applies to all aftersales products fitted to the car. This three-year warranty has a limit of 100 000 kilometres. In the case of aftersales products sold, the warranty covers a period of one year including the cost of fitting. Nissan estimates that 1 per cent of its aftersales products are compatible with other car producers' models.

Distribution policies

Nissan's aftersales outlets are geographically distributed among seven German states (Figure 8.19). In the federal state Schleswig-Holstein, Nissan has seventy outlets; in Lower Saxony, ninety; in Nord-Rhein-Westfallen, 150; in Hessen, 80; the Rheinland-Pfalz and Saarland federal states have combined 80 outlets; Baden-Württemberg has 100 outlets, Baveria 100; and the new federal states in the former German Democratic Republic have 250 outlets. These outlets are supplied both through rail and truck. Eighty per cent of the supplies are by truck and 20 per cent of the distribution of aftersales products takes place by rail. Nissan has a central store or Master Depot in Neuss supported by five mini-stores in Luckau-Berlin, Hamburg, Würzburg, Augsburg, and Essen.

Normal supplies take place weekly. Express service takes roughly sixteen

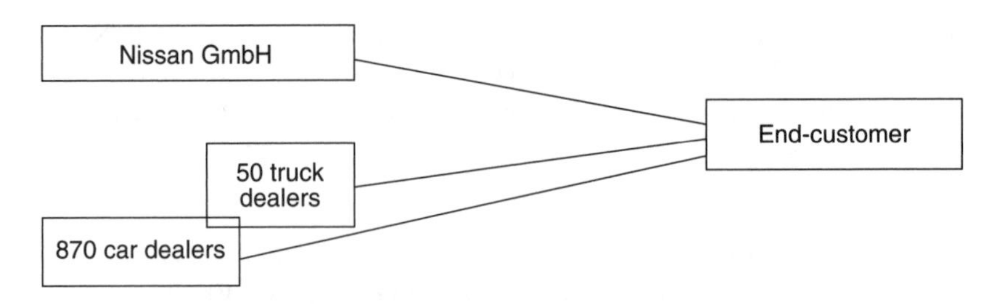

Figure 8.19 *Nissan's aftersales outlets*

hours and costs 50 per cent more than normal orders. Although the cost of normal deliveries is borne by Nissan, express delivery is split fifty–fifty with the dealers. The dealer pays 50 per cent of the costs and Nissan bears 50 per cent. In essence, Nissan is prepared to supply its dealer with normal deliveries and most probably as an encouragement to better dealer planning charges a cost for express orders. Nissan estimates the chances of a mistake occurring in its delivery system as 0.01 per cent and if such a mistake should occur the cost of correcting the mistake would be borne by Nissan. The distribution or ordering system is 95 per cent computerized with approximately 5 per cent still manual. Nissan does not have a recommended minimum stock level for its dealers.

Promotion policies

Nissan offers its dealers or outlets promotional support. This is made up of a brochure box, flyers, menu pricing, service leaflets, price lists, accessories brochures, posters and an electronic catalogue system, among others. Thirty per cent of the point-of-sale information is computerized and 50 per cent of the information contains estimated product fitting times designed to assist the dealer with better estimate fitting times.

Twenty per cent of the information contains product advantages that could be used as sales propaganda. At least about 30 per cent of the information regularly sent always includes up-to-date prices and 5 per cent contain legal regulations that affect and guide the use of the product. Only 5 per cent of the information contains very specific technical details. Dealer information is normally updated at least monthly with a general information check taking place quarterly. There is a recommended aftersales presentation concept for the outlets but this concept, though designed to assist the dealers, is not binding. Nissan is one of the few producers that has specialized aftersales advertisement campaigns in addition to new car advertisements. These promotional efforts are at present mostly dealer or regional based. However, the dealer must meet 50 per cent of the costs of the dealer-based advertisements with Nissan paying 50 per cent. In addition, Nissan employs seasonal offers to support its promotional efforts.

Service and related policies

Nissan does not offer its dealers/outlets assistance in the conception of aftersales strategies. Dealers are given a free hand to determine which path they intend to follow. The policy is to treat dealers as partner companies with the freedom to decide direction and market strategies. This policy looks set to change, according to *Autohaus* [57]. Nissan field representatives are currently undergoing training to enable them to offer management consultancy services to dealers.

Approximately 300 seminars or aftersales training blocks are offered yearly and, according to Nissan officials, these training blocks have a 100 per cent attendance rate. The training mostly takes place in Neuss. Areas of aftersales which the training addresses include technical know-how and customer contact methods, among others. In addition, Nissan has information lines open to its dealers between the hours of 9 am and 5 pm daily to offer assistance for aftersales problems. Furthermore, with its pan-European 24-hour service, Nissan offers its customers direct help in cases of emergency all over Europe. This service is linked to a dealer system in which certain dealers offer 24-hour service possibilities. At the moment there are eighty such service stations in Germany. Nissan also offers a service feature called its Nissan-Truepass 'Super plus'. This service is directed mainly at the owners of new Nissan cars and offers, among other things, assistance when a defect occurs more than 50 kilometres away from the customer's home.

Price policies

Nissan does not have special aftersales financing schemes. Dealers, however, receive discounts for products they buy. The type and form of discount is based on the products supplied and the volume the dealer buys. Although Nissan officials did not give the exact percentages of discount granted in an article in the *Unternehmer* magazine (May 1994), Nissan discounts were estimated at between 17 per cent and 50 per cent for parts and 24 per cent and 46 per cent for accessories normal orders [58]. For express orders, the article further estimated between 17 per cent and 30 per cent for parts and from 16 per cent to 35 per cent for accessories depending on the discount group to which the dealer belonged. The average discount for normal orders was estimated at between 40 per cent and 45 per cent for parts. Express parts receive a discount of between 22 per cent and 26 per cent. For accessories, the average was estimated at approximately 25 per cent for normal orders and 22 per cent for express orders.

At the moment Nissan ranks its payment conditions as not very competitive. No special bonus plans exist but there are plans to introduce a system that would be based on dealer attainment rates of pre-agreed targets. Ninety-nine per cent of the products are priced singly with 1 per cent priced as complete packages. Nissan recommends an hourly payment rate for its dealers which stands at DM90 per hour and varies according to region and dealer. At market prices, a general check-up costs the end customer approximately DM300. The average change of parts cost DM150 and accessories DM1080.

Environmental policies

Nissan has a recycling concept in practice. The objectives of the concept is to actively develop environmentally friendly products and to introduce a uniform concept for recycling all aftersales products all over Germany. According to reports in *Autohaus* [59], Nissan and six other Japanese importers recently signed a cooperation agreement with Preussag Recycling GmbH. Under this agreement known as MARI, Preussag undertakes to recycle the importers' old cars. At present 70–75 per cent of all aftersales products are recyclable. Nissan does not, however, bear the costs for the recycling – this is borne either by the dealer or the end-customer.

2 Policy effectiveness

Financial Results

The financial results of Nissan Germany are not published but, according to Nissan officials, aftersales turnover for the financial year end in 1994 was DM309 million. Parts accounted for approximately 7 per cent of this figure. Without fast-movers, they accounted for only 2 per cent. Fast-moving parts, therefore, are approximately 5 per cent of Nissan's turnover. Accessories, on the other hand, accounted for approximately 3 per cent of the 1994 financial year's turnover.

Ranking of policies

Nissan ranked itself, in relation to competitors, as partly effective with product policies, partly effective with promotion policies, distribution policies as very effective, price, environmental and service policies also as partly effective. Justification for its ranking is what it regards as the great potential for improvement in marketing and dealer aspects of its aftersales service. Logistics was ranked as very effective because the policies practised in distribution had ensured that Nissan was able to maintain the very good average standards found among most producers (Figure 8.20).

8.7.7 Toyota GmbH

1 Enablers and policies

General company information

Toyota is the third largest car producer in the world. It had, as at 1995, forty-eight production plants distributed all over the five continents. Established in

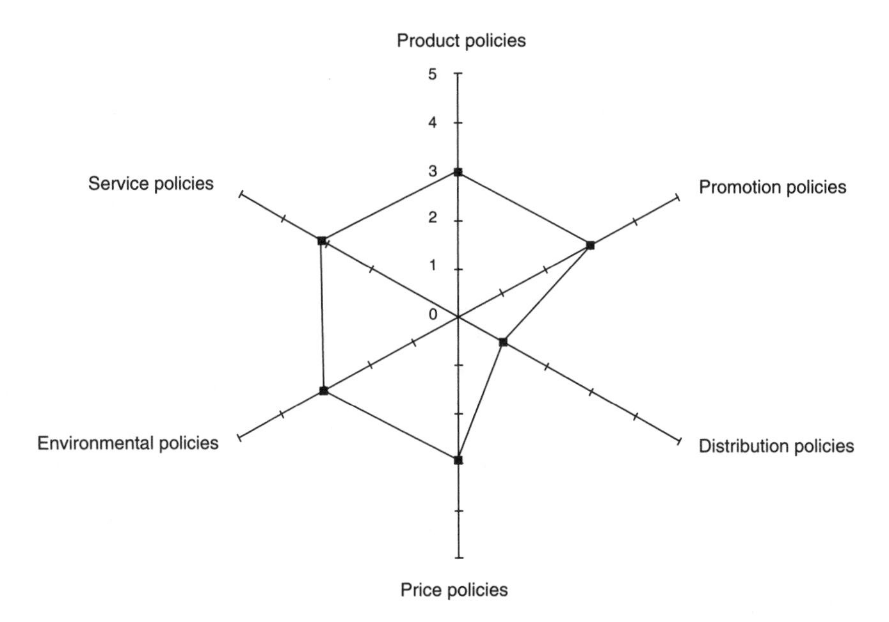

Figure 8.20 *Nissan's ranking of its aftersales policies*

1937 as the Toyoda Motor Corporation (the 'd' was later changed to a 't'), the Toyota group is the fifth leading car importer–producer in Germany. As at the end of 1994, its market share was 2.5 per cent. Toyota's aftersales strategy was summarized in its Toyota Fair concept 'to discover the wishes and requirements of the customer and to attempt to satisfy these wishes and requirements'. Toyota has a dealer network of 799 dealers or aftersales outlets in Germany divided into 634 direct dealers, 141 partner dealers and twenty-four branch dealers.

Toyota describes its aftersales objectives as customer satisfaction, customer retention, efficiency in work and job flow for both the distributor and dealer and high service absorption for dealers. Its major target groups are all Toyota customers. Although an aftersales department does not exist at Toyota, after-sales activities are taken care of by the service marketing department, parts marketing department, customer service and four regions with a total of twenty-four regional staff. According to Toyota sources, thirty-three people are directly responsible for these aftersales activities. Nine of these thirty-three coordinate from the offices in Cologne and twenty-four are field representatives. According to the 1994 annual report, Toyota employs a total of 793 workers.

Product policies

Toyota offers 150 000 single aftersales products broken down into three major groups: key products, the original parts group and accessories. The number of products belonging to the first two groups was not supplied but approximately 5000 products belong to the accessories group. According to Toyota sources, Toyota stocks all aftersales products that their models could need. Products are delivered directly from the Toyota Motor Company Japan through Toyota Europe.

All aftersales products carry the brand mark Toyota. While Toyota would classify some of its products as fast-movers, its major working classification is closely tied to an inventory system. In addition to this inventory system, it also maintains a demand classification system for each individual dealer. It offers a warranty on all its products. For parts the warranty is 12 months and for some accessories it could extend up to 36 months. The warranty covers cost of fitting. Certain Toyota products are compatible with other producer models but Toyota says it is unaware of the percentage of its products that are compatible.

Distribution policies

Toyota dealers receive their supplies from a central store in Cologne supported by three other stores in its regional offices in Hannover, Maisach near Munich and Bruchsal near Karlsruhe. The central store in Cologne stocks about 125 000 single products. Trucks are used for both normal and express deliveries and normal deliveries take place daily. This Just-in-Time concept of dealer delivery is unique. Express deliveries take a maximum of twelve hours. Both express and normal delivery costs are split 50 per cent paid by the dealer and 50 per cent paid by Toyota. Express therefore is at no extra cost except the normal 50 per cent cost to the dealer.

Toyota estimates the chance of a delivery mistake occurring to be 7 per cent and when such a mistake occurs, the cost of correcting the error is borne by Toyota. The distribution system is computer based including the dealer ordering system. Toyota has no recommended minimum stock levels for its dealers but has maximum stock levels for the dealers, depending on their markets. This is most probably due to the daily delivery system practised by Toyota.

Promotion policies

Toyota supports its dealers in their promotional activities. Direct support given includes technical information, product information, price lists through computer microfiches, aftersales brochures for each car model, advertising support, special complete price offers including parts and service, extra incentives, temporary price reductions and advertising for selected products including use of aftersales merchandising. The aftersales support, according to Toyota, is 100 per cent computer-based. Dealers are given a recommended aftersales presentation concept but the concept is, however, not binding in all outlets.

Toyota uses special aftersales advertisements as distinct from new car advertisements. These advertisements are, however, mainly regional and dealer-based. Payment for the cost of the advertisements varies, and some advertisement costs are shared between dealers and Toyota. Certain advertisements are paid for fully by Toyota and in some cases the dealer assumes 100 per cent cost responsibility. Other promotional measures employed by Toyota include seasonal offers, event marketing, sponsoring and public relations.

Service and related policies

Toyota offers full support in the conception of dealer aftersales strategies. These services cover the full range of all dealer business activities and are provided by the regional team of twenty-four (Figure 8.21). The dealer must, however, bear the cost of the services should any extra costs arise. In addition, Toyota offers a special end-customer programme called the Toyota Fair consisting of five direct end-customer services:

1 Toyota service through the dealers
2 Toyota Euro-care
3 Toyota quality through original parts
4 Toyota security through warranties and
5 The Toyota Key-Care service.

Furthermore, Toyota also offers aftersales training programmes. The volume of programmes to be offered is decided yearly and the dealer bears the cost of participating. Training takes place 30 per cent of the time at Toyota Germany's headquarters in Cologne and 70 per cent in the regions. Training mostly addresses technical know-how, customer contact, special aftersales marketing techniques and themes in the marketing of specific aftersales products. An information line is available between the hours of 9 am and 4 pm for all dealer aftersales problems and information.

Price policies

Toyota does not offer any end-customer-directed special aftersales financing schemes. However, through the Toyota credit bank, dealers can apply for and receive financial support. In addition, dealers are given discounts on the aftersales products ordered, the values of the discounts being based on a dealer grouping system. This system is made up of seven dealer groups but information on their exact make-up was, however, not revealed. The discounts offered fall broadly between 15 per cent and 50 per cent, the average discount being approximately 35 per cent. Furthermore, the dealers are also offered bonuses, their value ranging between 2 per cent and 10 of certain products purchased

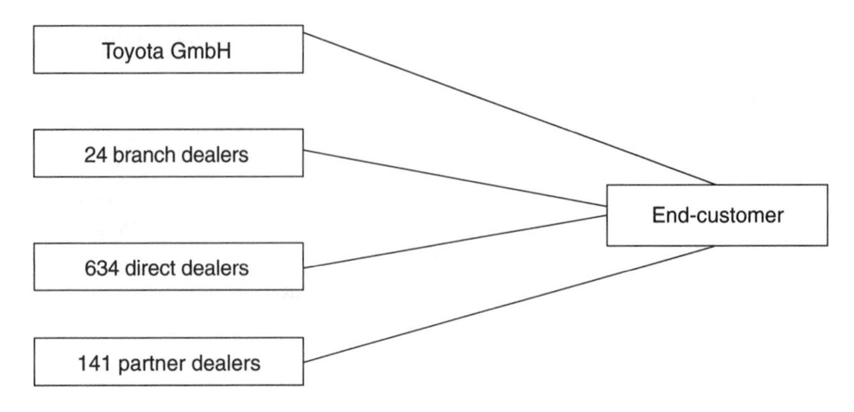

Figure 8.21 *Toyota's aftersales outlets*

from Toyota. Products used for the bonus calculation are simply described as 'key products'.

The majority of the aftersales products are priced as single products. However, some are priced as complete packages. Toyota does not recommend an hourly service payment rate to its dealers and considers its price policies as very competitive. According to *Autohaus* 1995 reports [50], Toyota recently reduced the prices of a wide range of aftersales products by as much as 30 per cent. The ADAC estimates the cost of aftersales maintenance for a Toyota Corolla 1.4 Xli to be about DM187 monthly. [60]

Environmental policies

According to Toyota sources, Toyota has a recycling concept in place. A major initiative covered by this concept is the recycling of bumpers. Apart from the bumper recycling, however, the dealer takes care of environmentally friendly disposal of wastes. Along with Nissan, Toyota is also party to the MARI cooperation contract with Preussag. There are plans to improve on the present environmental policies and Toyota has commissioned an internal study to examine new policy possibilities.

2 Policy effectiveness

Financial results

According to Toyota's annual report for the financial year ended December 1994, aftersales turnover was DM316 million. However, it is impossible to separate the products that are accounted for by the initial fittings (*erstausrustung*) from those sold through dealers.

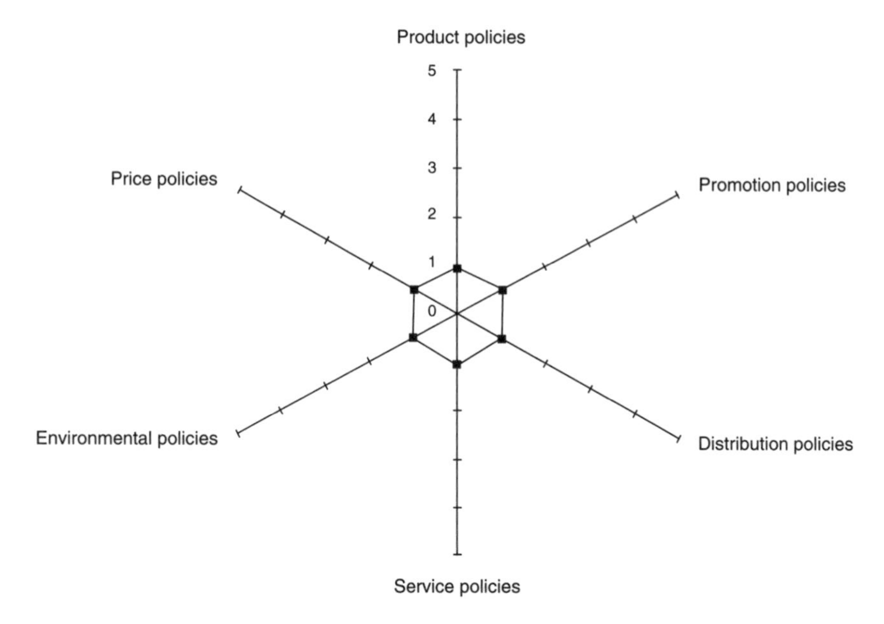

Figure 8.22 *Toyota's ranking of its aftersales policies*

Ranking of policies

Toyota ranked itself in relation to its competitors as follows: promotion, product, distribution, service, price and environmental policies very effective (Figure 8.22).

8.7.8 Ford Werke AG

1 Enablers and policies

General company information

Established on 2 October 1930, the former Ford Motor Company (renamed Ford Werke AG in 1939) is today one of the biggest car producers worldwide. As at 1995, approximately 10 million Ford cars had been built in Cologne Germany, with a market share of approximately 10.3 per cent (as at 1 July 1993). Ford is a major player in the German market [61] and its aftersales network is the largest of the producers/importers examined in this study (Figure 8.23), with 924 main dealers and a further 1518 lower-level dealers attached to this network. However, theoretically the lower-level dealers are contractually attached only to the main dealers. In addition, a number of independents offer aftersales service for Ford models.

Ford describes its aftersales objectives as 'to make Ford the number one aftermarket operation by providing owners of Ford vehicles an ownership

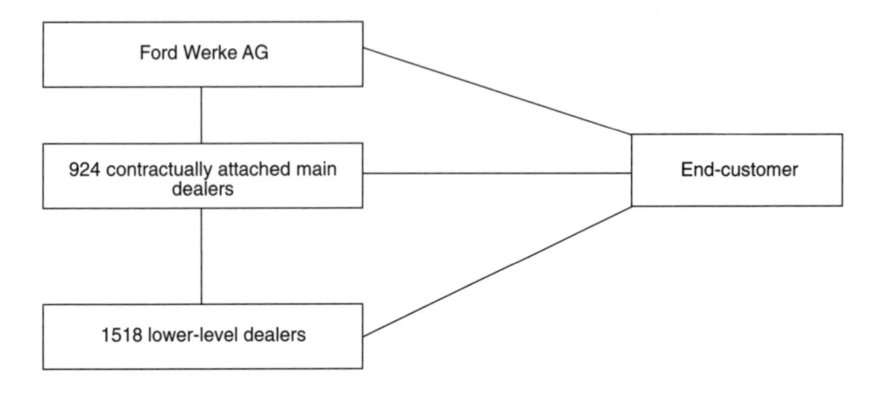

Figure 8.23 *Ford's aftersales outlets*

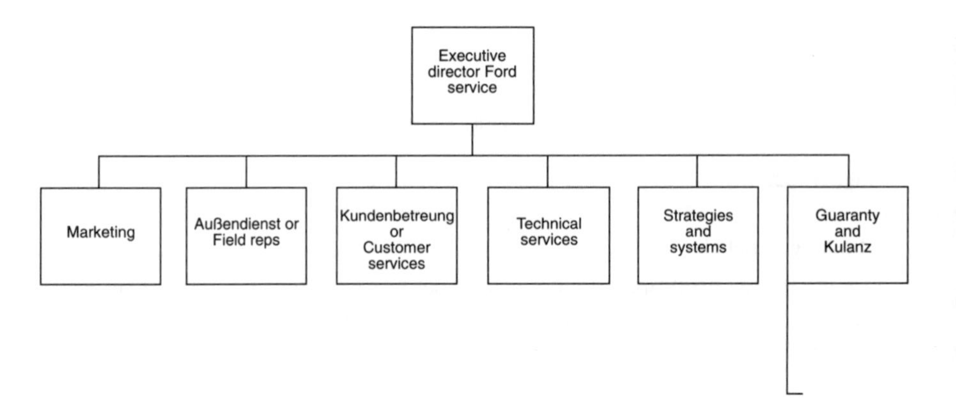

Figure 8.24 *Ford's organization chart*

experience that is so good [that] they naturally turn to Ford for vehicle purchases and aftersales needs'. These objectives are mainly targeted at all Ford owners. Aftersales at Ford is the responsibility of an executive director service. This department is further sub-divided into marketing, outside representatives, customer service, technical service, strategies and systems and guaranty and Kulanz departments (Figure 8.24). One hundred and sixty-eight workers are employed by the aftersales departments. Of these, one hundred and six work at the Cologne offices internally and fifty-two are assigned field responsibilities.

Product policies

Ford stocks 140 000 single aftersales products grouped into routine service, extended service, wear and tear, major repair, body and trim and accessories.

The number of products belonging to each group was, however, not specified. According to Ford sources, all aftersales products its models could require are stocked and 60 per cent of these products are manufactured by original equipment manufacturers (OEM) under contract by Ford. Those products manufactured under contract carry both the brand name of the manufacturer concerned and that of Ford. Of the 140 000 single products stocked, 500 can be classified as fast-movers (approximately 50 per cent of sales).

All aftersales products at Ford come with a one-year unlimited mileage warranty inclusive of cost of fitting where necessary. Approximately 5 per cent of Ford's aftersales products are compatible with other car producers' models.

Distribution policies

Ford's dealers are geographically coordinated through five districts: Hamburg with 154 main dealers and 259 lower-level dealers, Berlin with 249 main dealers and 270 lower-level dealers, Cologne with 159 main dealers and 169 lower-level dealers, Frankfurt with 183 main dealers and 380 lower-level dealers and Munich with 179 main dealers and 440 lower-level dealers. The outlets receive their product supplies by truck from the central store in Cologne or three other supporting stores. These supporting stores are also located in Cologne. Normal product orders are delivered weekly or within a five-day frame with the cost of delivery borne by Ford. Express delivery takes approximately 24 hours and the cost of delivery is borne by the dealer concerned. The chance of a delivery mistake occurring was estimated at 'far below 1 per cent'. Where such a mistake occurs, the cost of correction would be borne by Ford. Ford's product ordering system is fully computerized and dealer stock orders are processed and received by computer systems. However, dealers do not have to adhere to a minimum or maximum reorder level.

Promotion policies

Ford offers its aftersales outlets full promotional and point-of-sale support. Support information offered includes technical information, price lists, parts catalogue, repair manuals, computer system integrated support, among others. All data passed on to the dealers include specific information on product fitting times and up-to-date prices. Only some of the support information contains product specific advantages useful as sales propaganda. According to Ford, promotional and support information sent to the outlets is updated daily.

In addition, Ford offers its outlets a recommended aftersales presentation concept which, however, is not binding on all dealers. Both regional and dealer-based advertisements are used in reaching the target audience. However, there is no specific national aftersales advertisement. Under the slogan 'Ford die tun was' (Ford does something), Ford has a national campaign aimed at strengthening the corporate image. The costs of the regional or dealer-based advertisements are borne by the dealers. In addition to advertisements, Ford also employs seasonal offers, events, sponsoring and public relations activities as promotional measures for its dealers or outlets.

Service and related policies

Ford offers its dealers a range of service measures designed to further strengthen their aftersales programmes. Field representatives are engaged in on-site consultancy services and, where necessary, external expertise is used in the conception of dealer aftersales strategies. The cost of the external consultancy services is split between Ford and the dealers. The dealer pays between 60 per cent or 80 per cent of the external costs depending on the type required, while Ford pays between 20 per cent and 40 per cent.

In addition, dealers are offered aftersales training programmes based on a specialized aftersales training concept which address technical know-how for the workshop personnel, customer contact for the dealer service workers and management-related concepts for effective dealer business management. Each dealer has about five days training assigned for each of his service employees each year. On the whole, Ford offers 150 non-technical training programmes and 1250 technical seminars each year, and, according to Ford, dealers attend all the offered courses. Training takes place in specialized training schools with the cost of participation as 80 per cent paid by Ford and 20 per cent by the dealer. Furthermore, telephone contact lines are open between the hours of 7.30 am and 5 pm on working days where information and help on dealer problems is offered.

Price policies

Ford does not offer special aftersales financing schemes for the customer but dealer support is available in special cases. Dealers are offered product discounts which are both product and volume related and fall between 25 per cent and 55 per cent. In addition, dealers are offered a bonus programme which is tied to agreed and specified objectives and volume. All Ford's aftersales products are priced as single products and dealers are not given any recommended hourly payment rates. Ford describes its aftersales price policies as competitive but declines to give reasons. The cost for the end customer for selected aftersales services would vary based on individual dealer hourly payment rates. However, Ford estimates these rates to be between DM80 and DM110 per hour.

Environmental policies

Ford is a signatory to the International Chamber of Commerce's 'Charter for Sustainable Development' which sets out sixteen basic principles of environmental responsibility to which the signatories must adhere [62]. As early as 1993, (the first car producer to do so) Ford had developed recycling guidelines supported by practical product development guidelines aimed at facilitating easier and environmentally responsible disposal of products. Approximately 90 per cent of products are recyclable depending on the model.

Aftersales recycling activities already cover, among others, bumpers and batteries as well as product packaging. The costs of the disposal are, however, borne by the dealers except for package recycling, which is borne by Ford. In addition, Ford undertakes the recycling into the manufacturing process of engineers, transmission parts, cylinder-heads, injection pumps, radiators, starters, alternators, distributors, EEC modules, clutches, etc. This is at present done through a network of 100 certified competent recycling centres. The central thrust of the environmental policies at Ford is an attempt to establish, as early as the design phase, a basis for the recycling of the developed products. To this end there are clear guidelines aimed at integrating environmental considerations into the organizational activities. Ford also maintains a recycling database.

2 Policy effectiveness

Financial results

According to Ford's Annual Report for the financial year ended 31 December 1994, worldwide turnover of aftersales was placed at an aggregated sum of DM4.8 billion. Ford sources, however, estimated that 89 per cent of aftersales turnover comes from parts including fast-movers and fast-movers make up 50 per cent of this figure. Accessories account for only 11 per cent of the aftersales turnover.

Ranking of policies

Ford ranked its product policies as very effective, its promotion policies as effective and its distribution policies and price policies as partly effective. Environmental policies were ranked as effective and service policies were ranked as very effective (Figure 8.25). Justifications for these rankings were, however, not given.

8.7.9 Best practice in car aftersales service

Introduction

From a theoretical viewpoint, the description of specific practices as 'best' requires that a desired outcome or desired outcomes be specified. In essence, the description 'best practice' would then refer to practices which are best geared to meet specified goals.

However, the nature of organizational goals and objectives is complex and usually organizations face multiple and sometimes conflicting objectives. The resolution of these conflicts impacts on the policies pursued and an analysis of

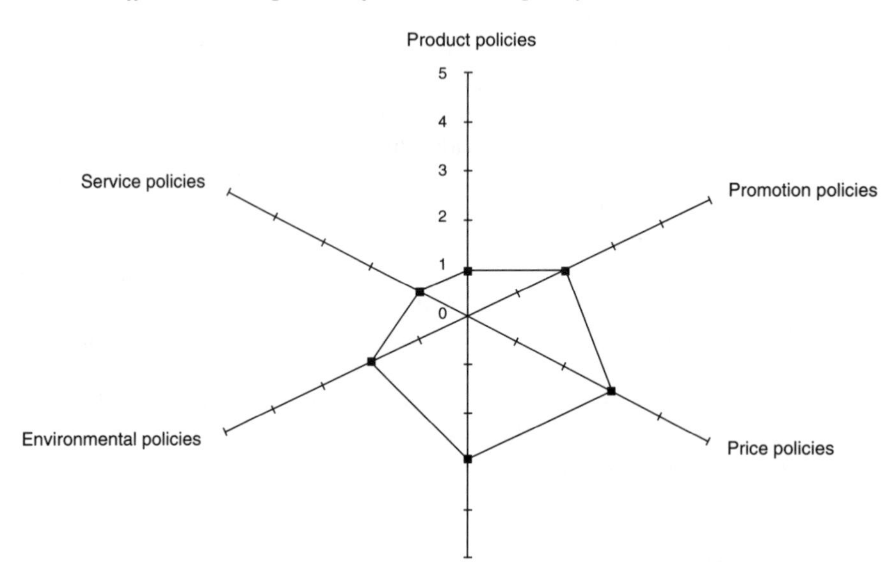

Figure 8.25 *Ford's ranking of its aftersales policies*

the effects and complexities of this problem is beyond the scope of this study. However, it is important to point out that the aftersales targets pursued by the organizations surveyed varied. It has not been possible to determine the extent to which these goals conflict with one another and this study did not set out to do so. It might, therefore, be inappropriate to present a list of practices best geared to meeting one goal without considering the effect of other goals.

What, therefore, has been done is to examine the practices set against a background of current market developments and problems; in essence, an examination or presentation of strengths and weaknesses using the major goal of customer satisfaction. This has been chosen because modern management research has shown that essential long-term organizational survival is the requirement that an organization pursue actively the satisfaction of its customers. For a definition of customer, the Bradford total quality management model of the customer–supplier process chain has been applied. This application shows that the dealer is the immediate customer in the aftersales process and it could, therefore, be appropriate to use the dealer's requirements in measuring the producer's policies. As a result of limitation of time and resources it was not possible to carry out a primary survey of the 5907 dealers contractually attached to the producers examined. However, interviews with some dealers and existing secondary sources (i.e. expert interviews and published material) has been relied upon. The nature of the car industry is dynamic, therefore an attempt has also been made to reflect this dynamism in the analysis. With these limitations and guidelines in mind, the next sub-section examines the aftersales policies of the producers using identified expectations or requirements of the customer in the process, the dealer, and the published opinions of experts and industry observers.

Product policies

Car producers manufacture 23 per cent of the aftersales products for the car industry, and 77 per cent is produced by the components industry [63]. Of this 77 per cent approximately 43 per cent is produced on contract by original component manufacturers for the car producers (Figure 8.26). As a result of this fact and the effective use of bonus, discounts and promotion schemes, producers have been able to achieve a market share of between 56 per cent and 77 per cent of all aftersales products (European market shares). Producers were also able to reduce competition from the component manufacturers both through the above policies and their original parts policies as well as by the use of legal product patents. This may account for the relatively large product inventories many producers maintain despite the fact that the majority of the turnover is accounted for by fast-movers.

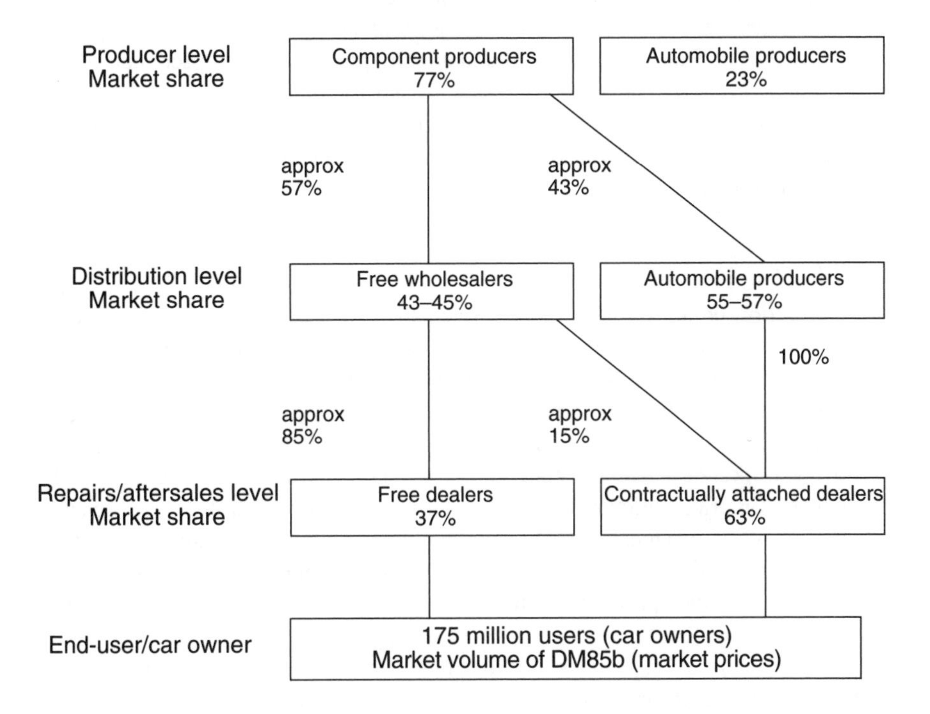

Figure 8.26 *Distribution of aftersales product market shares in Europe*

With the amendments to the GVO (the European law regulating the vertical marketing system) it is questionable whether it is in the interests of producers to continue to maintain large product inventories. Only 30 per cent of the fast-movers are manufactured by Fiat and a large number of its aftersales products are contracted out to original component manufacturers. Nissan, in contrast, has only 10 per cent of its products contracted out and has an inventory of 33 000 products in Germany of which 95 per cent of its fast-movers are manufactured internally. Toyota stocks approximately 120 000 single aftersales products and Ford stocks 140 000 of which only 40 per cent are manufactured by Ford, with 500 of these products being classified as fast-movers.

Based on inventory cost considerations and recent developments, Nissan and Fiat appear to have better policies although it must be borne in mind that the market size (i.e. the number of cars a producer has in circulation) and distribution policies (as is the case with Toyota's Just-in-Time distribution policies) does affect the number of products a producer stocks. However, perhaps far more important would be the impact of the amended GVO on the product policies and the market. With dealers free to source their products wherever they choose, competition is bound to increase. How this competition would be decided would perhaps depend to a large extent on the ability of the competing parties to manipulate price policies. Lower inventory levels in the coming turbulent times might prove to be an advantage. Other aspects of product policies appear to be universal and equally practised by all producers (e.g. warranty offered). The grouping system used for the products has not been examined because this aspect is closely tied to the internal processes and an examination would require much more in-depth quantitative analysis of the internal processes.

Distribution policies

Distribution policies appear to be universally rated as excellent in the industry. With a reported failure below 8 per cent the industry and indeed all the examined producers have been able to evolve effective distribution policies. Worth closer evaluation, however, is the Just-in-Time policy practised by Toyota and Fiat's policy of outsourcing logistics. Toyota's service level is, however, likely to have associated costs and a danger may lie in the effects of the costs on price policies. The unavailability of product cost details makes an analysis of the producer policies of either charging freight or supplying without freight charges difficult. In itself freight-free supply does not necessarily guarantee cheaper products as the cost of freight could easily be built into product costs.

Under distribution policies, it is also important to consider the relative number of outlets each producer has. This is closely influenced by the percentage of the producer's cars in circulation and perhaps the length of presence in the

German market. Ford by virtue of its longer market presence and larger market share is likely to have more dealers than the others. Experts have often maintained that the German market is overcrowded with dealers and a reduction is necessary to ensure continued dealer profitability. Using recent statistics [64] the car-to-dealer ratios of the four producers has been examined. With a total of 1796 dealers and 1 432 753 cars in circulation the Fiat group has a ratio of approximately 798 cars per dealer. This ratio is based on the assumption that Fiat dealers are able to achieve 100 per cent market share of the aftersales requirements of Fiat models. With contractually attached dealers having a market share of approximately 55 per cent, this implies with optimum dealer distribution a market potential of 438 cars per dealer on average. Nissan, with a total representation of 870 dealers and 1 014 205 cars, has a ratio of 1166 cars per dealer. Adjusted to reflect a 55 per cent market share this implies with optimum dealer distribution a market potential of 641 cars per dealer on average.

Toyota has a representation of 799 dealers and a total of 961 046 cars in circulation. Its ratio is 1202 cars per dealer adjusted to a market share of 55 per cent, which amounts to an average of 661 cars per dealer. Ford has 2442 dealers and a total of 4 052 439 cars. This is a car-to-dealer ratio of 1659 discounted to a 55 per cent market share and amounts to an average of 913 cars per dealer (statistics of cars in circulation are based on a date of 1 July 1994). In the absence of other comparative figures as to what an optimal dealer distribution should be, Ford's network appears to offer the dealers more aftersales potential on average.

Promotion policies

The best policies without effective communication both to the dealers and end-customers are likely to remain ineffective. Ford, with its national campaign 'wir tun was' ('we do something') and Fiat's 'patto chairo' ('we keep our word') initiatives are steps in the right direction. All the producers examined have been able to put some form of computer-based communication procedure in place, which has increased the frequency with which information is available to dealers, and Fiat needs to increase its computer-based initiatives in this respect. Worth mention is the 'corporate identity' schemes of some of the producers which are geared towards improving the image of the dealer outlets and developing a coherent organizational identity. These policies have been heavily criticized as they have come at a time when dealers, because of reduced profitability, are unwilling to take on extra investments. Some of Nissan's network 2000 measures are geared to improve their corporate identity.

The major measures used by the producers examined event marketing, public relations, dealer-based and regional advertising appears largely standard across the four subjects. Fiat's policy, by virtue of directly addressing the dealer problem and image, appears to be best in class. However, what aftersales also direly needs is a nationalized campaign geared towards informing customers of the quality of the producers and their networks.

Service and related policies

The service offerings of the producers examined reflect two major policy directions. The first is geared towards the dealer, (training programmes and consultancy services), the second towards the end-customer, (e.g. Nissan's truepas superplus, Toyota's Toyota Fair concept, among others). Of these geared directly to assist dealers, training is universal to all the producers examined. Dealer consultancy is, however, only a feature of Ford, Toyota and Fiat's policies. According to reports in the *Autohaus* magazine [57], Nissan would likely add this feature by March 1996. The use of external expertise by Ford in addition to its field representatives is an extra dimension of its dealer consultancy services. An advantage of this policy might be the ability to harness useful external techniques and solutions. Fiat's ISO 9002 quality certification process and its business management scheme are also good aspects of service policies.

Price policies

The most controversial of the producers' policies is price. In a market where it has become increasingly difficult to differentiate, the price instrument becomes even more important. The amendments to the GVO might have further increased the significance of this policy dimension as they break with the tradition in the marketing system in which producers could bind their dealers to a large extent to buy and use only their products.

Dealers are now allowed to decide where to buy or source their products as long as the quality is established to be the same as that of the producer. They could then, as it stands, decide to buy from the free market which is said to be often cheaper. According to Meinig [65], producers add as much as a 400 per cent increase on the prices they receive products from the component manufacturers. This increase is said to represent a profit margin of about 80 per cent. The major price policies of the producers examined in this study consisted of bonus and discount incentives and an attempt to encourage the dealers to purchase and use original parts. A consideration of the bonus or discount levels achieved for aftersales products by the producers is, in the light of the above argument, therefore secondary to true improvement.

It would be necessary for all producers to concentrate on achieving price levels that would be able to match the impending competition from the free market. Possible dealer reactions to the GVO amendments can be drawn from the reports in *Autohaus* of the formation of a 'buying cooperative company' by

Fiat dealers. This company is to be responsible for purchases of aftersales products, among others, from other European Union countries. Fiat dealers complain that the European parallel market's cheaper products in comparison to Fiat Germany has been damaging for their businesses [66]. In May 1995 Toyota was reported to have reduced prices for a number of its aftersales products [59].

Environmental policies

Active environmental protection has become a very important aspect of business in the 1990s. Aftersales policies must include comprehensive measures designed to ensure responsible and environmentally friendly activities. Ford's recycling concept with its associated recycling centres is a step in the right direction. It is likely that the future European Union laws and indeed consumer preferences would continue to emphasize environmental responsibility. There is, therefore, a need to develop a concept geared towards the environmental requirements of the aftersales process, and recycling of aftersales products is just one of the possibilities. Continuous environmental audits and environmental quality assurance are other possibilities. The present performance of the producers examined can definitely be improved.

8.7.10 Conclusions and recommendations

According to a survey carried out by *Autohaus* in 1993, approximately 59 per cent of a car dealer's yearly profits come from the aftersales service (Figure 8.27). The aftersales market, apart from being profitable, is also less subject to economic cyclic ups and downs [68]. With the increasing competitive pressures and structural changes taking place in the car industry, aftersales management must be moved to the forefront. Continuous process re-engineering and improvements need to become a feature of all players' policies and processes to ensure that aftersales 'rhetoric' is backed with optimal policies.

A major aftersales problem is the perception by the end-customer of the contractually attached dealers as being expensive. A number of factors are responsible for the creation of this image. However, this can be partly combated by policies aimed at informing the end-customer coupled with process improvements aimed at cost reductions.

Furthermore, producers' banks offering financing schemes for new cars might also consider offering the end-customer financing for aftersales products. The experience of major retailers and the success of mail order companies is based, among others, on the possibility that the customer has of paying in instalments. Clearly, this incentive could serve as a loyalty-retention scheme as the customer is bound for the credit period.

Aftersales improvements can only take place with a realization that future

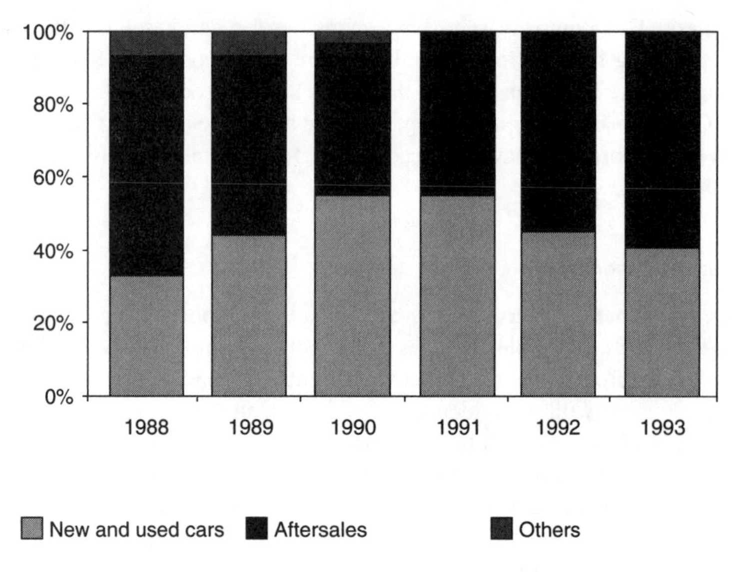

Figure 8.27 *Profitability of business units as a percentage of total profits. Source: adapted from Brachat [67]*

customer satisfaction would be strongly influenced by the quality of the aftersales delivery. The offerings in product design, technology and price are becoming increasingly difficult to differentiate. Aftersales presents differentiation potential that producers can use to strengthen their operations. However, the difference between the existence of potential and the actual achievement of competitive advantage must be taken into consideration [69].

Potential can only be converted to competitive advantage if the organization is able to convince customers to place equal subjective value on that aspect of the organization's offerings. The amendments to the European laws governing the vertical marketing system and the trends highlighted in this study are likely to result in a number of policy adjustments in the near future. The concept of benchmarking can assist in this coming re-engineering process. News reports of the activities of car producers already point to subtle policy redirection. For the organizations that are truly interested, this period offers the perfect opportunity to introduce radical changes.

8.8 The learning organization: results of a benchmarking study

8.8.1 Securing tomorrow through learning

Most if not all CEOs talk about business imperatives for the future and building a better tomorrow for their organizations. This is perhaps at a time when they are placing increasing emphasis on short-term results and getting the financials right. Is there therefore a paradox? Do they mean what they say? What is their real vision?

Perhaps to answer the above questions one could refer to an inquiry conducted by the Royal Society for the Encouragement of Arts. [70] The inquiry, made up of senior managers representing twenty-five top companies, concluded that:

- There is an overall gap in performance at a world-class standard because UK companies suffer from complacency and ignorance of world standards
- A national culture which is very adversarial
- There is an over-reliance on financial measures of performance and short-term focus.

The RSA report concludes that in order to create tomorrow's organization there has to be a radical shift through the creation of a more 'inclusive' approach and taking a stakeholder perspective. In particular, the report stresses the following:

- The need to define purpose, values and their effective communication at all levels
- Using the purpose, values, etc. as a baseline for designing individually tailored formulas for success and a meaningful framework for performance measurement
- Organizations need to recognize the importance of reciprocal relationships, and **learning** from all those who have a stake in the business in order to compete effectively
- Building close partnerships with customers and suppliers.

The report stresses the importance of learning in order to create tomorrow's companies. For instance, by specifically referring to the role of leadership and focusing on people:

- **Business leadership** includes 'defining, discussing, measuring and reporting on success in more inclusive ways' [71].
- **People** realizing the creativity and learning potential of all with whom the company has contact, not just employees. 'Participating in exploring the future of work, on the basis that successful companies can only flourish in a successful society'. 'rising tide lifts all boats.'

It is apparent therefore that creating tomorrow's organization requires a fundamental shift in attitudes towards people as the major asset and investing in the development of means of creativity, innovation and high business impact.

Jack Welsh, CEO and Chairman of General Electric, introduced the philosophy of 'change before you have to' to create a global organization, which is fit for competing in the 1990s. Since he joined GE in 1981, this change master embarked on a programme which stretched GE to become a global competitor, lean and mean, with annual revenues of $60 billion. He believes that it is the role of leaders to create a climate which constantly reminds people that change is a continuous process. Some of Jack Welsh's famous words include: **boundarylessness, speed, stretch**. According to Welsh, 'change should not be an event but rather a continuous process in the quest for success' [72]. Visionary leaders are referred to in a variety of ways such as:

- Change agents
- Transformational leaders
- Corporate revolutionaries
- Change masters.

8.8.2 Learning is synonymous with change management

The best way to describe learning is perhaps through the impact this can have on changing things, processes and organizations in their entirety. If effective change management is going to happen, organizations have to manage a variety of activities through direct and indirect means of encouraging learning. Wood [72] suggests a very useful list, which was observed in the effective management of change in various organizations such as Rank-Hovis:

- Corporate change through a top-down, bottom-up and middle-out approach relies on changing people's mind sets (how they think) and the culture of work (how they behave). This is the only way of creating ownership and commitment of the strategic task and also on relying on people's creative contributions
- There is a need to stress the importance of developing skills both at process management level and for the purpose of driving change and managing its implications. This can happen through various means such as teamwork and dynamic visual interaction
- Change can only happen through the development of robust processes and eliminating functional barriers in order to secure effective performance
- A specific role for managers and supervisors is to become change facilitative leaders, and develop everyone around them through coaching, guiding, mentoring and the encouragement of continual personal learning experiences
- Change has to be managed and cannot be left to occur in an *ad hoc* fashion. When change is introduced it has to be closely related to strategic planning and the competive arena outside. Change for the sake of change is worse

than no change at all. Change has to be considered as a means with ends, the end being effective competitiveness
- It is imperative for managers to link in employee's individual learning agendas to the business priorities and the core competencies required for the organizations concerned to achieve effectively their agenda in the marketplace. This link is so vital as it enables individuals to plan and manage their own learning tasks with the view of helping the accomplishment of the corporate goal and making a significant contribution.

As Wood argues:

> Above all, the change process is about people, and about unleashing their innate human potential to be the best they can be. If this is not recognized as the most fundamental principle of all, bottom-up and middle-out change lose their energy and meet with major barriers to change, resulting in eventual breakdown of the change programme.

The following is a list of critical factors for creating effective change [73]:

- Having a clear vision, leadership and communication and an effective decision-making process at all levels, driven by business imperatives
- It is important to have overwhelming support for change, otherwise if there are pockets of resistance these can hamper progress and advancement. One effective way of securing commitment and support is to encourage access to decision-making processes
- Building trust through openness and selling change as an opportunity for better competitiveness rather as a threat to job security. It is important to link change to learning, empowerment and risk taking and selling it as an opportunity for individuals' strengths and expertise to new roles to be assumed for delivering corporate goals and achieving the desired transformational state
- Having a change management policy which is well communicated and supported by all
- Focus and keeping the momentum going is crucial in continuing to empower people and get their full commitment
- Having champions, advocates and enthusiasts for change who are going to propagate the need and benefits at all levels within the organization. This is to be done through coaching and mentoring, for instance
- Tracking and monitoring success is critical to ensure that change is effective and leading to the desired outcomes
- Using external catalysts for help, assistance, support and inspiration.

8.8.3 Benchmarking learning

Since learning is an all-encompassing approach, an adaptation of a model proposed by Carr [74] was used to assess corporate learning within Unilever plc and Allied Domecq plc. Both are operating in the fast-moving consumer goods sector and have more similarities than dissimilarities.

Unilever plc

Unilever's aims are to effectively meet the needs of the global consumer in its specific markets. Its strength is in branded and packaged consumer goods, in foods, detergents and personal products. Unilever has over 1000 strong and successful brands being marketed worldwide. Some of the strengths of Unilever are in the following areas:

- Understanding the needs of customers and consumers
- Product innovation
- Investment in R&D
- Creative and effective marketing
- Expertise in manufacturing technology
- People.

As far as people are concerned, and this is the area scrutinized in this section, in a corporate document issued in 1992 the following statement was written: 'Unilever recognizes that its employees are the lifeblood of the business; it is their skills and commitment which determine success.'

Allied Domecq plc:

Allied Domecq is a world leader in spirits and retailing. Spirits and wine account for 62 per cent of Allied Domecq's trading profit (1995 figures) and retailing for 26 per cent. They have additional areas which include brewing and food manufacturing. Like Unilever, Allied Domecq has very strong brands such as Ballantine's, Beefeater, Kahlua, Courvoisier and Teacher's. In retailing, they have 13 800 retail outlets, comprising 4100 pubs, 1500 off-licences, and 8200 franchised stores with brands such as Victoria Wines, Big Steak Pubs, Mr Q's. There is a determination at Allied Domecq to drive through the strengths of its brands, as in the case of Unilever. This is expressed by its Chairman (Michael Jackaman) in the 1995 company report, as follows:

> Our consumer brands are all-important to our future growth. We focus on them intently, giving them consistent and imaginative support. The quality of our brands and the speed of our service are key determinants of success.

Table 8.11 *Background to data collection*

Elements of comparison	*Unilever*	*Allied Domecq*
Number of respondents	14	12
Number of functions represented	10	12
Number of businesses represented	08	04

8.8.4 The survey instrument

The survey instrument was based on a questionnaire proposed by Carr [74] structured in six areas, covering the following:

1 Core qualities for creativity
2 Key characteristics of a creative organizational system
3 Measures of organizational flexibility
4 Measuring diversity and conflict
5 Developing creative organizational systems
6 Developing effective and successful teams.

Ten functions were represented within Unilever plc and twelve in Allied Domecq. Responses were received from eight Unilever Businesses and four from Allied Domecq. In all, twenty-six questionnaires were received out of a total 100 sent (Table 8.11). The results are discussed in the following sections, under each of the six elements used and based on Carr's model.

8.8.5 The seven core qualities of a creative organization

In this area which covers core elements of being a learning organization and operating in total quality management approach, Unilever overall has a better profile than Allied Domecq. The gaps appear to be in three specific areas:

- Within Unilever, there is a better approach to people's contributions
- Within Unilever, there is a better approach to problem solving
- There is a more *ad hoc*, trial and error approach within Allied Domecq than within Unilever (Figure 8.28).

8.8.6 The basic characteristics of a creative organizational system

This area examines how both Unilever and Allied Domecq put learning into practice. Both organizations appear to be lacking in many areas, in particular:

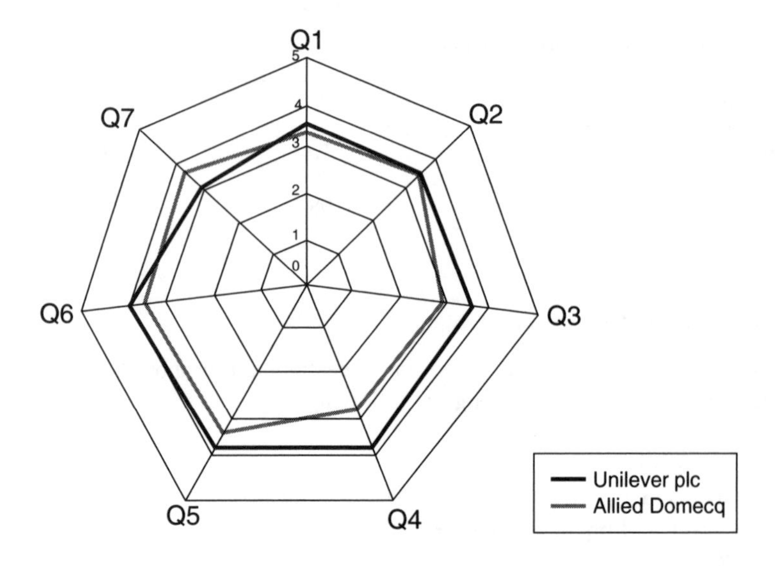

Figure 8.28 *The seven core qualities of a creative organization*

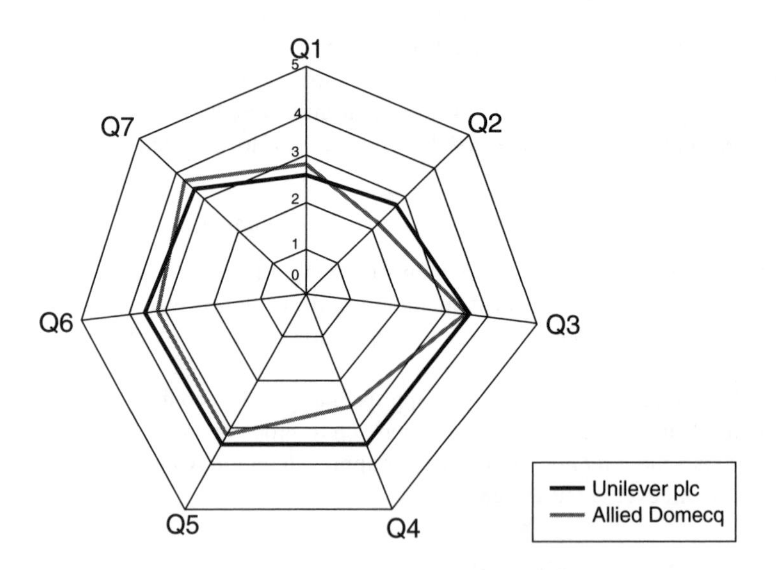

Figure 8.29 *The basic characteristics of a creative organizational system*

- There is concern overall the allocation of resources for problem solving
- There is not a true culture of problem solving in both organizations and using problems for injecting in new learning
- The evidence about pioneering approaches and innovativeness in relation to other organizations is not particularly strong (Figure 8.30).

8.8.7 The ten attributes of a flexible organization

This area attempts to map whether there is clear evidence of a positive culture of total involvement, a proactive approach to managing change and whether the notion of true empowerment is really visible. Both profiles once more do not appear to be at a world-class level. In particular, the following areas appear to be of concern:

- It appears that there is not a positive climate for creativity and an approach which encourages people to experiment and learn
- The notion of empowerment is not widely recognized and there is still concern over hierarchies and rigid structures (Figure 8.31).

8.8.8 Using diversity and conflict

There is much concern about diversity and how conflict is considered. Both profiles once again do not appear to reflect best practices in the development

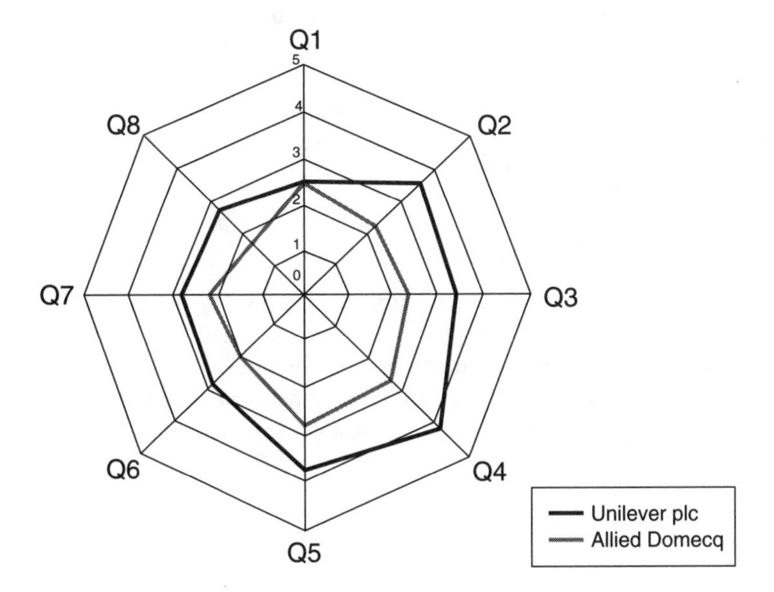

Figure 8.30 *The eight rules for a creative HRM system*

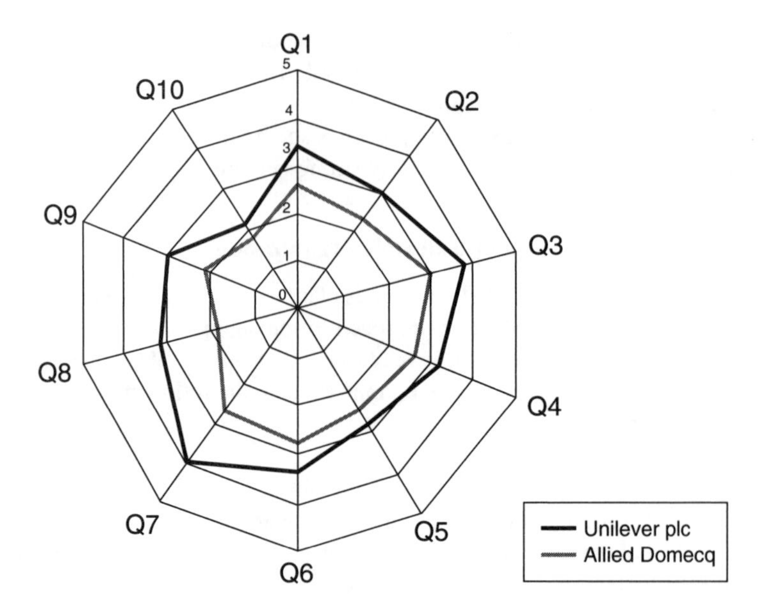

Figure 8.31 *The ten attributes of a flexible organization*

of a corporate climate where goals and targets are effectively communicated and closely linked to individual development needs and where opportunities for promotion, participation in the decision-making processes are not widely abundant. In particular, the following areas need to be highlighted:

- Opportunity to apply for positions at all levels
- Individual goals and needs are not integrated effectively into the structure of organizations (Figure 8.29).

8.8.9 The eight rules for a creative HRM system

This is again one of the most critical areas for creating the learning organization. It is very much about people choice, selection and development, how they are motivated and rewarded and how levels of synergy are obtained by encouraging team work. Although Unilever appears to be much better than Allied Domecq in this respect, nonetheless, apart from one or two areas, there are many of concern:

Positive

- Effective training and its link to functions
- The quality of people hired (creative potential)

- Tasks and projects are made to be the key motivator for people
- People are hired for keeps and not just to fill in a short-term need.

Areas for improvement

- The ways HRM systems are developed and put into practice are not really about harnessing creativity and maximizing the full potential of people and how they work together
- The reward and recognition system is purely and simply monetary and not enough emphasis is placed on creative contributions
- The incentive systems are not put together to support the creative performance of people (Figure 8.32).

8.8.10 The nine characteristics of successful teams

This is the last area examined and specifically relates to aspects of team work, goal clarity, culture of supporting teams, whether values and guiding principles are shared and how the goal-deployment process concentrates on the work of teams (Figure 8.33). Once again, Unilever appears to have a slightly better profile than Allied Domecq. The general indication is that this is not really an area of great concern, and although there is no one single practice which highlights world-class status, nonetheless, the effort deployed indicates

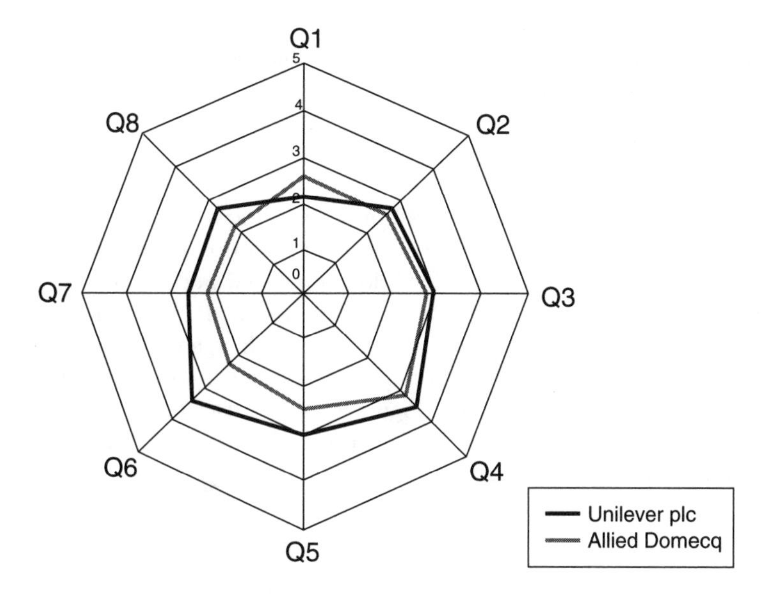

Figure 8.32 *Using diversity and conflict*

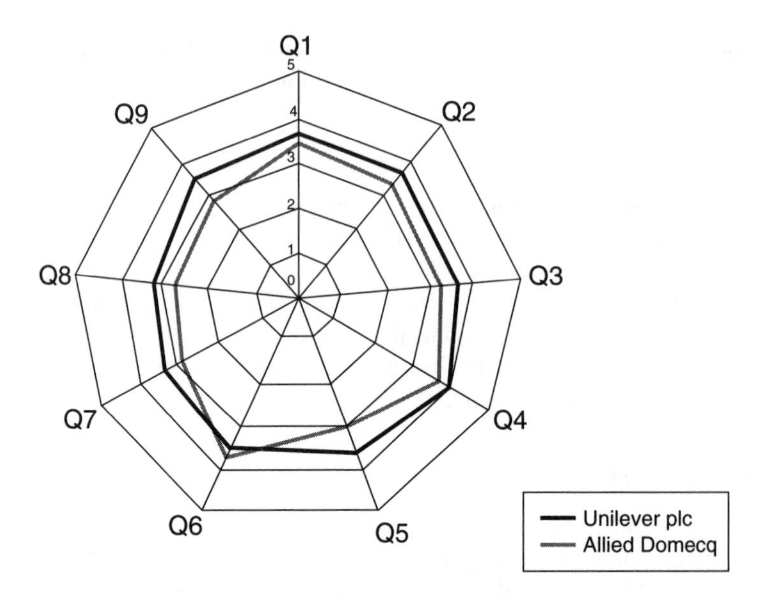

Figure 8.33 *The nine characteristics of successful teams*

above-average practices. Perhaps areas which need to be highlighted and would require improvement cover:

- Renewals for team performance. There is still a strong focus on individuals
- There is clear recognition that individuals within the team work context do add value and efforts are recognized.

8.8.11 Conclusions

This brief benchmarking exercise indicates that Unilever has the edge over Allied Domecq and perhaps is more advanced in its effort to creating a learning organization. The efforts, however, do not indicate world-class status and there are many deficiencies which may need attention. When plotting the aggregated responses of all the respondents from each organization for each of the six parameters examined, Unilever appears to be much stronger than Allied Domecq, as Figure 8.34 illustrates: Figure 8.35 illustrates the gaps in performance between Unilever and Allied Domecq under the six key elements of developing a learning organization.

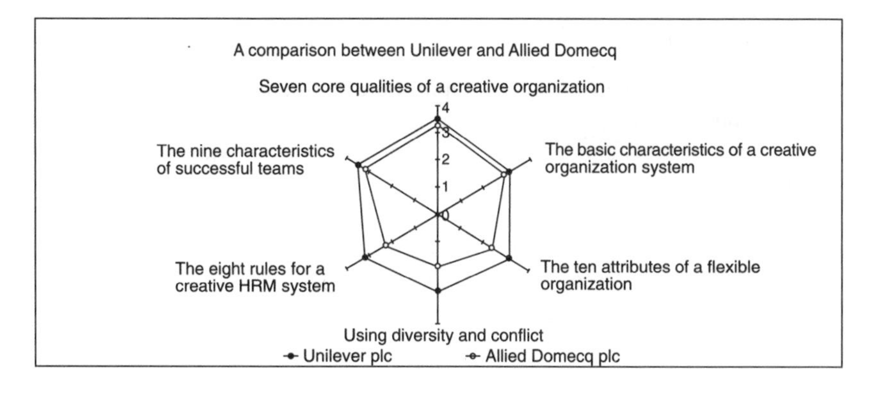

Figure 8.34 *Benchmarking learning in FMCG*

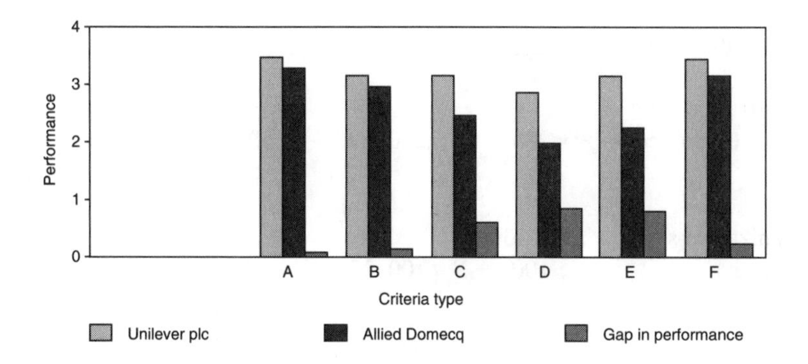

Figure 8.35 *Benchmarking learning: gaps in performance*

8.9 Measuring creativity in organizations (adapted from Carr [74])

A diagnostic tool for measuring creativity in organizations

1 Core qualities for creativity
2 Key characteristics of a creative organizational system
3 Measures of organizational flexibility
4 Measuring diversity and conflict
5 Developing creative HRM systems
6 Developing effective and successful teams

Scoring process

Please place a tick against the most appropriate box in the context of your own organization.

The same process applies to all the questions and the various statements given:

☐	0–2	Not true at all
☐	2–4	Not true
☐	4–6	Some truth
☐	6–8	True
☐	8–10	Very true

Management function: ______________________________

Organization name: ______________________________

Organization size: (No. of employees)	<250	250–500	500–1000	>1000
	☐	☐	☐	☐

Organization size: (turnover)	£ million >100	100–250	250–500	<500
	☐	☐	☐	☐

A. The seven core qualities of a creative organization

	0–2 ☐ Not true at all	2–4 ☐ Not true	4–6 ☐ Some truth	6–8 ☐ True	8–10 ☐ Very true
1 My organization intends to be creative, and its members are creative	☐	☐	☐	☐	☐
2 The members of my organization direct their creativity toward goals that are important to the organization	☐	☐	☐	☐	☐
3 The members of my organization perform at a very high level	☐	☐	☐	☐	☐
4 My organization is good at identifying important problems and finding the opportunities within them	☐	☐	☐	☐	☐
5 My organization takes time to formulate a problem in depth before deciding how to solve it	☐	☐	☐	☐	☐
6 My organization considers a wide variety of alternatives before committing itself to a specific direction	☐	☐	☐	☐	☐
7 My organization often has to make frequent attempts, none of them quite satisfactory, before it comes up with the right solution	☐	☐	☐	☐	☐

B. The basic characteristics of a creative organizational system

	0–2 ☐ Not true at all	2–4 ☐ Not true	4–6 ☐ Some truth	6–8 ☐ True	8–10 ☐ Very true
1 My organization understands the relationship between outputs and waste quite differently from the way traditional organizations do	☐	☐	☐	☐	☐
2 My organization uses waste, surprise, and invisible consequences as a source of opportunity	☐	☐	☐	☐	☐
3 My organization makes those operating each system responsible for controlling it	☐	☐	☐	☐	☐
4 My organization provides the individuals who operate systems free access to the resources they need	☐	☐	☐	☐	☐
5 My organization ensures that each player works both to achieve the overall goals of the organization and to meet specific objectives that support these goals	☐	☐	☐	☐	☐
6 My organization provides directly to performers tactical feedback that they can use to improve operations	☐	☐	☐	☐	☐
7 My organization provides and uses feedback from the market at all levels, but particularly at the strategic level	☐	☐	☐	☐	☐

C. The ten attributes of a flexible organization

	0–2 ☐ Not true at all	2–4 ☐ Not true	4–6 ☐ Some truth	6–8 ☐ True	8–10 ☐ Very true
1 Our organization is built on a high level of trust	☐	☐	☐	☐	☐
2 We expect everyone in our organization to tell it like it is – and to ask the questions necessary to find out how it is	☐	☐	☐	☐	☐
3 We not only permit but encourage everyone to communicate with everyone else	☐	☐	☐	☐	☐
4 When a problem arises, we look for solutions, not scapegoats; we neither pistol-whip members for making mistakes nor excuse the mistakes	☐	☐	☐	☐	☐
5 We focus on problems and opportunities, not on personalities and power structures	☐	☐	☐	☐	☐
6 Our organization uses shared values, goals, and objectives to support and enhance self-management	☐	☐	☐	☐	☐
7 We include our customers and suppliers in our decision-making processes	☐	☐	☐	☐	☐
8 We are always scanning the horizon and proactively anticipating change; we are good at creating the future	☐	☐	☐	☐	☐
9 We promote ownership and entrepreneurship everywhere	☐	☐	☐	☐	☐
10 We encourage play, daydreaming, and even silliness in the organization	☐	☐	☐	☐	☐

D. Using diversity and conflict

	0–2 ☐ Not true at all	2–4 ☐ Not true	4–6 ☐ Some truth	6–8 ☐ True	8–10 ☐ Very true
1 Our organization has taken positive steps to ensure that all positions, including those at the very highest levels in our organization, are open to *everyone*	☐	☐	☐	☐	☐
2 We understand and actively further diversify without limiting it to a narrow multicultural understanding	☐	☐	☐	☐	☐
3 In policy, and in fact, we communicate to all our members at all levels that they are important players, essential to the success of the organization	☐	☐	☐	☐	☐
4 We ensure that everyone focuses on goals and the objectives required to achieve them, not on narrow requirements or procedures	☐	☐	☐	☐	☐
5 We accept the validity of the individual goals and needs of *all* our members and constantly structure the organization to meet them as fully as possible	☐	☐	☐	☐	☐
6 Our organization not only permits but encourages everyone to communicate with everyone else and insists that this occurs across functional, ethnic, and any other divisions	☐	☐	☐	☐	☐
7 We expect conflict to be helpful, expect it to be openly expressed, and train our members in effective conflict-resolution techniques	☐	☐	☐	☐	☐
8 We have an affirmative programme that empowers members of minority groups, women and the disabled to express their own goals and to work to achieve them	☐	☐	☐	☐	☐

E. The eight rules for a creative HRM system

	0–2 ☐ Not true at all	2–4 ☐ Not true	4–6 ☐ Some truth	6–8 ☐ True	8–10 ☐ Very true
1 Our organization fits all its human resources management practices together into a coherent system that supports constant creativity	☐	☐	☐	☐	☐
2 We constantly hire creative people	☐	☐	☐	☐	☐
3 We hire performers primarily to make a lifetime contribution, not just fill a specific job	☐	☐	☐	☐	☐
4 We train constantly and integrate the training into our ongoing operations	☐	☐	☐	☐	☐
5 We make the challenge and worth of the work itself the most important motivator for high performance – for everyone	☐	☐	☐	☐	☐
6 We are expert at balancing people and work, creative and routine work, and individual and team work	☐	☐	☐	☐	☐
7 We don't motivate by money, but our compensation system is congruent with our emphasis on creativity	☐	☐	☐	☐	☐
8 We ensure that all our organization's incentives fit together to support creative performance	☐	☐	☐	☐	☐

F. The nine characteristics of successful teams

	0–2 ☐ Not true at all	2–4 ☐ Not true	4–6 ☐ Some truth	6–8 ☐ True	8–10 ☐ Very true
1 Our organization (especially the managers responsible to implementing teams) and the team members have shared values that support team work	☐	☐	☐	☐	☐
2 Our teams have clear, worthwhile, compelling goals	☐	☐	☐	☐	☐
3 These goals can be accomplished only by a team	☐	☐	☐	☐	☐
4 There is a genuine need for every member of each team	☐	☐	☐	☐	☐
5 All members of the teams are committed to the goals of their teams	☐	☐	☐	☐	☐
6 Each team has specific, measurable objectives to accomplish as a team	☐	☐	☐	☐	☐
7 Each team gets effective feedback on its performance	☐	☐	☐	☐	☐
8 We have specific renewals for team performance, not just individual performance	☐	☐	☐	☐	☐
9 Members of the teams are competent both as individuals and as a team	☐	☐	☐	☐	☐

References

1 Stikker, A. (1992) Sustainability and business management, *Business Strategy and the Environment*, **1**, No. 3, 1–8.
2 Newman, J.C. and Breeden, K.M. (1992) Lessons from environmental leaders, *Columbia Journal of World Business*, Fall & Winter Issues, 210–21.
3 Polonsky, M. and Zeffare, R. (1992) Corporate environmental commitment in Australia: A sectoral comparison, *Business Strategy and the Environment*, **1**, Part 2.
4 Greeno, J.L. and Robinson, S.N. (1992) Rethinking corporate environmental management, *The Columbia Journal of World Business*, Fall & Winter, 222–32.
5 Barnett, S. (1992) Strategy and the environment, *The Columbia Journal of World Business*, Fall & Winter, 202–8.
6 Simon, F.L. (1992) Marketing green products in the triad, *The Columbia Journal of World Business*, Fall & Winter, 269–85.
7 Wescott II, W.F. (1992) Environmental technology cooperation – a *quid pro quo* for transnational corporations and developing countries, *The Columbia Journal of World Business*, Fall & Winter, 144–53.
8 Gray, R. and Owen, D. (1992) Environmental reporting award scheme, *Integrated Environmental Management*, No. 19, May, 8.
9 Booth, J.A. (1993) *An assessment of BS5750 within UK organisations*, MBA dissertation, University of Bradford Management Centre.
10 Barthelemy, J.L. (1993) *In search of best practice in the management of quality systems audits for continuous improvement: the case of ICL*, MBA dissertation, University of Bradford Management Centre.
11 Redman, N. (1993) Benchmarking to support a vision, *Proceedings* of Benchmarking for Strategic Advantage Conference, London, 6 July 1993, ICM Marketing Ltd.
12 Hudiburg, J.J. (1991) *Winning with Quality – the FPL Story*, Quality Resources, USA.
13 Saunders, M. (1993) Improving customer service – benchmarking what matters to the customer, *Proceedings* of Benchmarking for Strategic Advantage Conference, London, 6 July 1993, ICM Marketing Ltd.
14 Burt, D.N. (1989) Managing suppliers up to speed, *Harvard Business Review*, **67**, No. 4, 127–35.
15 Ellram, L.M. (1990) The suppliers selection decision in strategic partnerships, *Journal of Purchasing & Materials Management*, **26**, No. 4, 8–14.
16 Garwood, R.D. (1990) Partnerships: supplier beware! *Production & Inventory Management Review & APICS News*, **10**, No. 8, 28.
17 Ingman, L.C. (1991) Buying 'right': pushing quality upstream, *Pul & Paper (PUP)*, **65**, No. 4, 175–7.

18 Jick, T.D. (1990) Customer–supplier partnerships: human resources as bridge builders, *Human Resource Management*, **29**, No.4, 435–454.
19 Wrigglesworth, J. (1994) Keynote Address – The Future of Building Societies, Euro Forum Conference Papers.
20 Coles, A. (1993) Recent developments in building societies and the savings and mortgage markets, The Building Societies Association, London.
21 McKillop, D.G. and Ferguson, C. (1993) *Building Societies: Structure Performance and Change*, Graham & Trotman, London.
22 Bootle, R. (1993) Analysing likely economic trends in the 1990s and their likely impact on retail financial services, IIR Publication – Papers on customer competitiveness, May.
23 Morgan, G. and Sturdy, A. (1993) Bancassurance: innovating strategies in financial services, Services Industry Management Research Unit Conference Papers, April.
24 Nellis, J. (1994) Financial services in Europe, *Council of Mortgage Lenders bulletin*, March.
25 Llewelyn, D.T. (1995) Market imperfections and the target-instrument approach to financial services regulation, *The Services Industry Journal*, **15**, No. 2, 203–15.
26 Morgan, G. (1992) The globalisation of financial services: the European Community after 1992, *The Services Industry Journal*, **12**, No. 2, 192–209.
27 Altunbas, Y., Maude, D. and Molyneux, P. (1995) *Efficiency and mergers in the UK (Retail) Banking Market.*
28 McKillop, D.G. and Glass, C.J. (1994) A cost model of building societies as producers of mortgages and other financial products, *Journal of Business, Finance and Accounting*, **21**, No. 7, October 1031–46.
29 Aitken, J. (1995) Lloyds Bank: black horse wins gold cup, *UBS Global Research.*
30 Wilkinson, A., McCabe, D. and Knights, D. (1996) Looking for quality: a survey of quality initiatives in the financial services sector, *TQM Journal*, **7**.
31 Grant, R.M., Shani, R. and Krishnan, R. (1994) TQM's challenge to management theory and practice, *Sloan Management Review*, Winter, 25–35.
32(a) Hill, S. and Wilkinson, A. (1995) In search of TQM, *Employee Relations Journal*, Special Issue on TQM, **17**, No. 3.
32(b) Garvin, D.A. (1991) How the Baldrige award really works, *Harvard Business Review*, November/December, 80–93.
32(c) Milgrom, P. and Roberts, J. (1992) *Economics, Organisation, and Management*, Prentice Hall, Englewood Cliffs, NJ.
33 Kano, N. (1993) A perspective on quality activities in American firms, *California Management Review*, Spring, 12–31.
34 Easton, G.S. (1993) The 1993 state of US TQM: A Baldrige examiner's perspective, *California Management Review*, Spring, 32–55.

35 Juran, J.M. (1993) A renaissance with quality, *Harvard Business Review*, July/August, 42–53.
36 Greene, R. (1995) *Competent Re-engineering: advice, warnings, and recipes from eye witnesses*, Addison-Wesley, Reading, MA.
37 Witcher, B. (1993) *The Adoption of TQM in Scotland*, Durham University Business School.
38 Kearney, A.T. (1993) *Total Quality: Time to Take off the Rose Tinted Spectacles. A survey of a cross section of UK firms*, IFS Publications, London.
39 Economist Intelligence Unit (1992) *Making quality work – Lessons from Europe's leading companies*, EIU/Ashridge Management School report.
40 Zairi, M. (1992) *Competitive Benchmarking: An Executive Guide*, Technical Communications (Publishing) Ltd.
41 Dean, J. and Bowen, D. (1994) Management theory and total quality: improving research and practice through theory development, *Academy of Management Review*, **19**, No. 3, 392–418.
42 Black, S. (1994) *Total Quality Management: the critical success factors*, PhD thesis, University of Bradford.
43 Saraph, J.V., Benson, P.G. and Schroder, R.G. (1989) An instrument for measuring the critical factors of quality management, *Decision Sciences*, 810–29.
44 Bossink B.A.G., Geiskes, J.F.N. and Pas, T.N.M. (1992) Diagnosing Total Quality Management – parts 1 and 2, *Total Quality Management Journal*, **3**, No. 3, 223–31; **4**, No. 1, 5–12.
45 Porter, L.J. and Parker, A.J. (1992) Total Quality Management – the critical success factors, *Total Quality Management Journal* **4**, No. 1, 13–22.
46 Ramirez, C. and Loney, Y. (1993) Baldrige award winners identify the essentials of a successful quality process, *Quality Digest*, January, 38–40.
47 UBS (1994) *Building Societies Research: The Major Players*, Research Report UBS Phillips and Drew.
48 Longbottom, D. and Zairi, M. (1996) TQM in financial services: an empirical study of best practice, In Kanji, G. (ed.), *TQM in Action*, Chapman and Hall, London, pp. 242–54.
49 Kaplan, R.S. and Norton, D.P. (1992) The Balanced Business Scorecard – measures that drive performance, *Harvard Business Review*. January/February, 71–79.
50 Müller, W. and Reuss, H. (1995) Veränderungen Wettbewerblicher Erfolgsfaktoren im Automobilmarkt, in Müller, W. and Reuss, H. (eds), *Wettbewerbsvorteile im Automobilhandel Strategien und konzepte für ein erfolgreiches vertragshändler management*, Campus Verlag, Franfurt.
51 Motitz Spillen, E. (1995) Wettbewerbsvorteile durch eine aktive und systematische Kundenotientierury. In Müller, W. and Reuss, H. (eds), *Wettbewerbsvorteile im Automobilhandel Strategien und konzepte für ein*

erfolgreiches vertragshändler management, Campus Verlag, Frankfurt, p. 11.
52 BBE (1994) Branchen-report band 1 and 2 BBE Unternehmensberatung Köln, Cologne, 293.
53 Müller, W. (1995) Kundenzufriedenheit beginnt beim Hersteller, *Unternehmermagazin*, 3 May, 38.
54 Reuss, H., and Wolk, H. (1990), Distributions-szenario die revolution im Automobil-vertrieb: Konfliktmanagement im Automobilfranchising und wege seiner handhabung, unpublished BBE internal document, pp. 20–27.
55 Meinig, W. (1995) Dealer Satisfaction Index, based on the study by Eurotax Schwacke/Autohaus Verlag, 19 June.
56 Fütterer, D. (1994) Veränderungen in der europäischen Automobilindustrie angesichts der weltweiten ökonomischen Herausforderungen, in Meinig W., (ed.), *Wertschöpfungskette Automobilwirtschaft: Zulieferer-Hersteller-Handel, Internationaler Wettbewerb und Globale Herausforderungen*, Gabler Verlag, Wiesbaden p. 40.
57 *Autohaus* (1995) Network 2000, *Autohaus Magazine*, No. 16, August, 48.
58 Steiner, A. and Herrnberger, J. (1994) Kostenoptimierung im Ersatzteilwesen, *KFZ-Betrieb Unternehmermagazin*, 4 May.
59 *Autohaus* (1995) 'News briefs', *Autohaus Magazine*, No. 12, May.
60 *ADAC Motorwelt Magazine* (1995) April.
61 BBE (1994), Branchen-report band 1 and 2, BBE Unternehmensberatung Köln, Cologne, p. 115.
62 Brandstetter, W. (1994) Rank und Aufgaben der Umwelttechnik in der Automobilindustry, in Meinig W., (ed.), *Wertschöpfungskette Automobilwirtschaft: Zulieferer-Hersteller-Handel, Internationaler Wettbewerb und Globale Herausforderungen*, Gabler Verlag, Wiesbaden, p. 92.
63 *Autohaus Magazine*, No. 6, 20 March, 60.
64 Kraftfahr Bundesamt statistical records released for 1994 (German car statistical office).
65 Meinig, W. (ed.) (1994) *Wertschöpfungskette Automobilwirtschaft: Zulieferer-Hersteller-Handel, Internationaler Wettbewert und Globale Herausforderung*, Gabler Verlag, Wiesbaden.
66 *Autohaus Magazine* Number 16, 1995, page 50.
67 Brachat, H. (1995) Computergestützte Kommunikationssysteme in Handel und Kundendienst, in Müller, W. and Reuss, H. (eds). *Wettbewerbsvorteile im Automobilhandel Strategien und konzepte für ein erfolgreiches vertragshändler management*, Campus Verlag, Frankfurt, p. 242.
68 Eriksen, K. (1990) *Das Marktverhalten der Automobilhersteller auf den KFZ-Teilen-Märkten der Bundesrepublik Deutschland*, Göttingen Vandenhoeck & Ruprecht, p. 31.
69 Müller, W. and Reuss, H. (1995) Veränderungen Wettbewerblicher

Erfolgsfaktoren im Automobilmarkt, in Müller, W. and Reuss, H. (eds)., *Wettbewerbsvorteile im Automobilhandel Strategien und konzepte für ein erfolgreiches vertragshändler management*, Campus Verlag, Frankfurt, p. 23.
70 Tomorrow's Company: the role of business in a changing world, July 1995, The Royal Society for the Encouragement of Arts (RSA), London.
71 Tomorrow's company, *Quality of Working Life (QWL) News and Abstracts*, No. 124, 4, Autumn 1995, ACAS, London.
72 Salazar, R. (1995) 'Leading corporate transformation', *World Executive's Digest*, August, 10–12.
73 Wood, R. (1995) New articipative series on people, work and change, *The Strategic Planning Society News*, November, 4–5.
74 Carr, C. (1994) *The Competitive Power of Constant Creativity*, AMACOM, American Management Association, New York.

Further reading

Antcliffe, D. (1996) Texas Instruments Europe: strategies for excellence, in Kanji, G (ed.), *TQM in Action*, Chapman & Hall, London.

Balmer, J. and Wilkinson, A. (1991) Building societies: change strategy and corporate identity, SIMRU Conference papers, April.

Birro, K. (1991) *The evolution of planning systems*, PhD thesis, University of Bradford.

Camp, R. (1989) *Benchmarking: The Search for Industry Best Practices that Lead to Superior Performance*. ASQC Quality Press, New York.

Coles, A. (1994) The Director General's Address, *Mortgage Finance Gazette*, Conference Issue, 10–14.

Crosby, P.B. (1979) *Quality is Free*, McGraw-Hill, New York.

Cruise O'Brien, R. and Voss, C. (1992) In search of quality, London Business School working paper.

Deming, W.E. (1986) *Out of the Crises*, MIT Centre for Advanced Engineering.

EFQM (1994) *Total Quality Management – The European Model for Self-Appraisal.*

Feigenbaurn, A.V. (1983) *Total Quality Control*, McGraw-Hill, New York.

Greene, R. (1993) *Global Quality*, ASQC Quality Press, New York.

Hammer, M. and Champy, J. (1993) *Re-engineering the Corporation*, HarperCollins, New York.

Juran, J.M. (1992) *Juran on Quality by Design*, The Free Press, New York.

MBNQA (1994) The Malcolm Baldrige National Quality Award: 1994 Criteria, National Institute of Standards and Technology.

Porter, L.J. and Oakland, J.S. (1993) Teamwork for mission achievement, ASQC Quality Congress 1993, 229–35.

Rank Xerox (1992) Rank Xerox European Quality Award submission document 1992, ASQC Quality Press, New York.
Sinclair, D. (1994) *Performance Measurement: An Empirical Study of Best Practice*, PhD thesis, University of Bradford 4.
Snape, E. and Wilkinson, A. (1991) Human resource management in building societies: making the transformation, SIMRU Conference papers, April.
Thwaites, D. (1991) *Innovation and Marketing Practices in UK Building Societies*, PhD thesis, University of Bradford.
Thwaites, D. and Edgett, S. (1991) Aspects of innovation in a turbulent market environment: empirical evidence from UK building societies, *The Services Industry Journal*, **11**, No. 3, July, 346–61.
Witcher, B. (1994) Clarifying Total Quality Management, working paper Durham University Business School.
Wrigglesworth, J. (1994) The mortgage and savings markets, *Mortgage Finance Gazette*, Conference issue, 29–31.
Zairi, M., Letza, S. and Oakland, J.S. (1994) Does TQM impact on bottom line results? *The TQM Magazine*, **6** (1), 38–43.

9 Bringing it all together: effective project management

The beautiful is as useful as the useful. Perhaps more so

Victor Hugo

The best is the enemy of the good

Voltaire

A thing of beauty is a joy forever

John Keats

9.1 Lever Brothers Ltd: a success story

Lever Development Centre, Port Sunlight, is part of Lever Brothers Ltd UK, belonging to Unilever plc. This success story started with William Hecketh Lever, the founding father of Unilever's Detergents and Personal Products Business. He progressed from a one shilling a week wage for cutting bars of soap in his father's grocery business at the age of 16 to an estate valued at £1.6 million at the time of his death in 1925 at the age of 73. The father of the Lever Brothers success story is known for quotes such as:

> If we leave the human factor out of our business calculations, we shall be wrong every time.
>
> There is a tradition that women know nothing about business but this has never been my experience.
>
> The man who makes no mistakes does not usually make anything.
>
> I realize that half the money I spend on advertising is wasted.
> The trouble is, no one can tell me which half.

Lever Brothers UK is now a full success story. It blends in expertise, high energy, excellence in brand names and a commitment to quality and continuous improvement. Using the 1995 baseline Lever Brothers Ltd enjoys a formidable competitive performance with the following facts:

- Turnover (1995): £460 million, 75 per cent home trade, 25 per cent export. Export sales have grown from £38 million in 1991 to £110 million in 1995
- Number of employees: 1560

- Brands: 15 major brands, 70 products. Lever has over 22 per cent value share of its markets in the UK
- Top five brands: Persil, Domestos, Comfort, Jif and Dove
- Advertising spend: approximately £45 million (1995)
- Main manufacturing sites: Port Sunlight (Soaps and Liquids). Warrington (Powders)
- Production volumes:

Production volumes:	UK	285 000 tonnes
	Europe	155 00 tonnes (across 15 countries)
	Rest of the world	75 000 tonnes (across nine countries)

  Total production – 515 000 tonnes
- 1995 capital investment: £113 million over the last five years (69 per cent ahead of statistical depreciation).

9.2 Lever Development Centre, Port Sunlight: the need for benchmarking

Lever Development Centre, Port Sunlight, is one of the various development centres based in Europe, dealing with fabric products. It is made up of six sections. Lever's primary reason for using benchmarking was born out various reasons, not all related to business imperatives. Although there was a concern over some innovations not being delivered effectively to the marketplace, benchmarking was introduced for 'Identifying and measuring what Lever Development Centre, Port Sunlight must do to become a world-class innovation centre'.

A team was set up in September 1994 with two primary objectives in mind:

1. To understand the workings/effectiveness of Lever Development Centre (LDC) Port Sunlight by examining a variety of projects current and past.
2. To compare the performance of the LDC with innovation processes in other companies, thus determining performance gaps and highlighting areas of best practice to be implemented through an effective action plan.

The first milestone led to a wide variety of benefits. Using a combined approach relying on workshops, project case histories, internal and external customer surveys, a thorough understanding of the effectiveness of LDC's core activities was achieved. Among the key strengths the project team identified were the following:

- Good team working within sections
- Quality of the personnel employed within LDC

Among the areas which were thought to show potential for improvement were:

- Quality of briefs given to project teams
- Broad appreciation of business strategy/consumer requirements, in the context of project management
- The process for resource allocation
- The type and quality of communication between various sections
- Quality of process documentation.

Using outcomes from the first stage, the benchmarking team set about to put together a plan for pursuing the external comparative perspective, which mainly, seeks to identify areas of best practice in project management related to innovation processes. This bold move was unprecedented in many ways since innovation and its management is considered to be the lifeblood of any business and therefore most organizations would guard very jealously and secretly their approach to innovation management and would, under normal circumstances, be very unwilling to participate in any benchmarking exercise, let alone share sensitive information.

9.3 Benchmarking LDC's innovation project management

9.3.1 What is benchmarking?

Identifying what to benchmark, which aspect of the innovation process to focus on and what is critical to the business proved to be a very challenging task. Apart from anything else, innovation project management is a very wide, encompassing area containing soft and hard aspects. Relying on information gathered from Stage 1 the team came up with a framework composed of seven key elements, namely:

- Structure of NPD/innovation process
- Resourcing
- Books of techniques
- Team working
- Documentation
- Measures
- Culture of innovation.

Table 9.1 illustrates the description and definitions for each of the seven critical areas used in the benchmarking exercise.

Table 9.1 *LDC processes for projects: definition of aspects of parameters being evaluated*

Benchmarking is the process of finding and implementing best practices. (It requires being humble enough to admit that one's own practices need to be improved and we can learn from others.) The LDC benchmarking team is benchmarking the development/innovation process using the following parameters:

1 *Development/innovation process structure*
How well defined and easy to communicate/understand/utilize is the structure being used?
How rigorously is it applied and controlled?

2 *Resourcing*
How effectively are projects prioritized, and the required cross-functional resources calculated, deployed and monitored?

3 *Tools and techniques*
How comprehensive is the range of techniques available for all aspects of the innovation process, and how effectively are they used?

4 *Team working*
What criteria are used to select team leaders and members and how is effective empowerment and cross-functional cooperation achieved. What training is given in project leadership and team working?

5 *Documentation*
How well documented and archived are the critical stages of the innovation process, so that the control stages (e.g. briefs, charters, CSFs at gates) are rigorously applied, checked and recorded? What specific systems are used?

6 *Measures*
What measures are made before, during and after the innovation process and how do they assist in improving the effectiveness of the process? What review mechanisms exist for the measures?

7 *Culture of innovation*
How and how well does the organization promote a culture of innovation and encourage all to contribute towards it by appropriate organizational devices and rewards?

9.3.2 Questionnaire development

This area was not treated very likely and using several workshops and an iterative process, a questionnaire was put together under the seven key headings. The full questionnaire is included in Appendix 9.1. Most of the questions were to be used as prompts and an *aide-mémoire* to the individuals concerned. The questionnaire did not represent a prescriptive way of conducting the interviews but rather was meant to bring about discipline and consistency in the approach to be used by the various team members.

9.3.3 Scoring system development

The development of performance indicators to be used for gap establishment and analysis was not an easy task. Several attempts had to be made before the finalization on an approach which was thought to be generically applicable and which would be capable of ensuring consistency in comparisons, thus giving the 'apples versus apples' outcome. The key learning point from this is perhaps that:

- Measurement is not a static but rather a dynamic process
- Measurement is a complex activity
- Measures are about individual language and specific cultures
- Measurement as an activity, only represents 'means to ends'.

Developing a generic approach which would bear in mind all the above factors was therefore the only way forward.

- As a starting point it was thought to be important enough to categorize and identify the degree of impact each of the seven parameters has on the effective management of innovation projects;
- Through a weighting approach of a high/medium/low scale and whether the project team members can have direct control/high degree of influence over each area and finally whether the need to carry out external benchmarking is an urgent one, a matrix was developed (Table 9.2).

Having established the different criticality levels the next stage was to produce the core components for each of the seven key parameters and allocate scores according to the degree of importance reflected by each component (Table 9.2).

As Table 9.3 indicates, a benchmarking total of 50 points (equivalent to 100 per cent) was to be used as the common method of scoring the various partners concerned (see also Table 9.4)

9.3.4 Partner selection

Subscribing to the guidance of a best-practice methodology, the team used the following approach to arrive at a comprehensive list of potential partners:

- Internal benchmarking – Are there any sister companies known to have effective innovation project management approaches?
- Externally – Who are the companies known to excel in their consistency of product launches and effective innovation management? (generic approach).

- Who are the best companies known to specifically excel at specific parameters of the seven to be examined? (functional approach).

In total, a list of nine companies was put together, representing two sister companies and seven best in class partners. Table 9.5 provides background information on each of the partners involved.

Table 9.2 *Weighting procedure applied to the seven parameters chosen for benchmarking of the innovation process*

Used three criteria, giving high, medium or low weighting:
- Fundamental importance to innovation process
- LDC is empowered to act in area
- Need to carry out benchmarking to make it happen

	Fundamental importance	*Within brief*	*Need BM to happen*	*Nett*	*Weight*
Process structure	H	L	M	**M**	**100**
Resourcing	H	H	H	**H**	**120**
Tools and techniques	M	M	M	**M**	**100**
Team working	H	M	L	**M**	**100**
Documentation	H	H	H	**H**	**150**
Measures	M	L	H	**M**	**100**
Culture of innovation	H	H	M	**H**	**150**

Table 9.3 *Weighted scoring system used to score aspects of the seven parameters chosen for benchmarking of the innovation process*

Parameter		*Score*	*LDC*	*Another*
Process structure:				
• Innovation		10		
• Overall system		20		
• Gate management		20		
• Control system		15		
• Idea generation		20		
• Time management		15		
	Sub-total	**100**		
Resourcing:				
• Prioritization		30		
• Allocation		30		
• Innovation proportion		30		
• Time allocation		30		
	Sub-total	**120**		

Parameter		*Score*	*LDC*	*Another*
Tools and techniques:				
• Methods		30		
• Designing in quality		30		
• Customer needs		30		
• Other quality tools		10		
	Sub-total	**100**		
Team working:				
• Definition		10		
• Selection		60		
• Decision taking		30		
	Sub-total	**100**		
Documentation:				
• Overall		60		
• At gates		50		
• Archiving		40		
	Sub-total	**150**		
Measures:				
• What		25		
• Where		25		
• How		25		
• Why		25		
	Sub-total	**100**		
Culture of innovation:				
• Incentives		50		
• Creativity		40		
• Structure		40		
• Patents		20		
	Sub-total	**150**		
	Benchmark total	**830**		
	Percentage	**100**		

Notes: Allocation of scores to companies using the above system is subjective. Bias was minimized by mixing team members on the visits and scoring relative to previous visits. Early scoring is essential while meeting fresh in mind but all low/high scores questioned before acceptance.

Table 9.4(a) *Benchmarking study record: weighted scoring of all parameters against BM partners*

Parameter	*Max. score*	*J*	*I*	*G*	*H*	*F*	*E*
Process structure							
• Innovation	10	5	5	9	–	8	5
• Overall system	20	15	15	18	18	17	17
• Gate management	20	5	20	20	15	12	20
• Control system	15	5	10	12	13	10	10
• Idea generation	20	5	12	16	18	15	10
• Time management	15	5	8	8	10	19	12
Sub-total	**100**	**40**	**70**	**83**	**74/90**	**72**	**74**
• Prioritization	30	20	22	25	23	20	25
• Allocation	30	15	20	25	25	15	25
• Innovation proportion	30	15	20	–	20	15	20
• Innovation strategy	30	10	10	20	25	20	25
• Time allocation	30	10	13	12	20	10	20
Sub-total	**150**	**70**	**85**	**82/120**	**113**	**80**	**115**
Tools and techniques							
• Methods	30	20	20	28	25	20	22
• Designing in quality	30	15	15	28	17	20	17
• Customer needs	30	15	20	28	28	25	15
• Other quality tools	10	0	2	5	5	6	5
Sub-total	**100**	**50**	**57**	**89**	**75**	**71**	**59**
Team working							
• Definition	10	5	5	8	5	6	6
• Selection	60	45	40	55	55	24	45
• Decision taking	30	20	20	20	25	20	20
Sub-total	**100**	**70**	**65**	**83**	**85**	**71**	**71**
• Overall	60	20	20	30	50	40	20
• At gates	50	25	40	40	35	35	35
• Archiving	40	15	18	17	20	25	15
Sub-total	**150**	**60**	**78**	**87**	**105**	**100**	**70**
Measures							
• What	25	5	10	18	20	15	18
• Where	25	5	10	18	20	12	18
• How	25	5	10	18	20	15	18
• Why	25	5	5	18	20	20	18
Sub-total	**100**	**20**	**35**	**72**	**80**	**62**	**72**

Parameter	*Max. score*	*J*	*I*	*G*	*H*	*F*	*E*
Culture of innovation							
• Incentives	50	0	10	30	45	5	0
• Creativity	40	0	10	10	35	35	15
• Structure	40	5	10	25	25	25	25
• Patents	20	0	10	–	0	–	15
Sub-total	**150**	**5**	**40**	**65/130**	**105**	**65/130**	**55**
Benchmark total	**850**	**315**	**430**	**561/800**	**637/840**	**521/830**	**516**
Percentage	**100**	**37**	**51**	**70**	**76**	**63**	**61**

Table 9.4(b) *Benchmarking study record: weighted scoring of all parameters against BM partners*

Parameter	*Max. score*	*D*	*C*	*B*	*A*	*Co. name*
Process structure						
• Innovation	10	9	7	5	9	
• Overall system	20	18	18	15	18	
• Gate management	20	18	18	10	20	
• Control system	15	14	13	8	15	
• Idea generation	20	18	15	5	18	
• Time management	15	12	13	8	11	
Sub-total	**100**	**89**	**84**	**51**	**91**	
Resourcing						
• Prioritization	30	25	25	20	25	
• Allocation	30	25	25	20	15	
• Innovation proportion	30	25	25	5	25	
• Innovation strategy	30	25	25	5	21	
• Time allocation	30	25	25	20	13	
Sub-total	**150**	**125**	**125**	**70**	**99**	
Tools and techniques						
• Methods	30	30	25	20	25	
• Designing in quality	30	28	25	5	20	
• Customer needs	30	28	20	10	28	
• Other quality tools	10	8	8	0	5	
Sub-total	**100**	**94**	**78**	**35**	**78**	

Parameter	*Max. score*	*D*	*C*	*B*	*A*	*Co. name*
Team working						
• Definition	10	8	9	5	6	
• Selection	60	55	55	50	50	
• Decision taking	30	25	28	15	20	
Sub-total	**100**	**88**	**92**	**70**	**76**	
Documentation						
• Overall	60	55	55	20	50	
• At gates	50	45	45	35	45	
• Archiving	40	35	30	15	35	
Sub-total	**150**	**135**	**130**	**70**	**130**	
Measures						
What	25	22	20	15	18	
Where	25	22	20	15	18	
How	25	22	20	15	18	
Why	25	22	20	15	18	
Sub-total	**100**	**88**	**80**	**60**	**72**	
Culture of innovation						
• Incentives	50	45	35	25	15	
• Creativity	40	35	20	10	15	
• Structure	40	35	20	20	15	
• Patents	20	15	5	10	18	
Sub-total	**150**	**130**	**80**	**65**	**63**	
Benchmark total	**850**	**749**	**669**	**421**	**609**	
Percentage	**100**	**88**	**79**	**50**	**72**	

9.3.5 Partner contacts

Each of the nine companies was contacted through a formal letter and a copy of the questionnaire to be used during the interviews. This was followed up by:

- Finalizing dates of visits
- Sending an agenda for the meeting
- Forwarding names of team members to be involved in each visit.

The LDC benchmarking team also prepared a standard presentation for the benefit of each of the partners involved in the study. This was thought to be important for a variety of reasons, with perhaps the opportunity of developing long-term links and giving an insight to each partner involved on where LDC

Table 9.5 *Brief description of the part of each company visited*

Elida-Fabergé Ltd A Unilever company, which for about the last 2 years has been putting into practice the Integral APP tool & techniques for the innovation process. In addition, they have general experience of QFD. The part of the business visited during the benchmarking study was concerned with the development of Deodorants and Male Toiletries, based in Leeds.

D2D (Design to Distribution Ltd) A subsidiary of ICL based in Stoke-on-Trent, which is concerned with production of components for the electronics industry, being involved in all stages from design to delivery of the finished product to their customers. Innovation of processes (e.g. speed and flexibility of reaction to orders) rather than producing new products is becoming increasingly important, with the aims of achieving high standards of quality and customer satisfaction.

Ford Motor Company Ltd Different aspects of motor car design and manufacture are dealt with by different groups. The group visited was concerned with power train development for the Small and Medium Vehicle Centre within the Research and Engineering Centre in Basildon, Essex, where the Ford Customer Satisfaction Process (based on QFD principles) is being put into practice.

Cadbury The visit was to the main factory at Bournville in Birmingham. The business is based on chocolate, confectionary and soft drinks and is currently introducing innovation process management systems. The Cadbury team represented a range of functions – planning, development, quality, packaging and marketing.

Birds Eye Wall's Ltd A Unilever company, which for the last 2 to 3 years has been putting into practice the Integral APP tools & techniques for the innovation process, and has an established commitment to TQ. The operation visited was designated as the Development and Quality Audit (Ice Cream) operation, a group within the ice cream factory in Gloucester.

Dow Corning (Europe) A specialist silicone manufacturer and supplier based in La Hulpe in Belgium. The benchmarking team consisted of personnel who had been heavily involved in introducing TQ innovation procedures into every aspect of the business over a number of years. The company is very customer focused and has processes in place for delivering innovations, from idea capture to product launch.

Rover Group The benchmarking study visit was to members of the Design and Engineering Centre at Gaydon, near Warwick. Several aspects of this operation were represented by the Rover team members, e.g. research, design, engineering, operations and finance.

BP Chemicals This business consists of three divisions – Polymers, Commodity Chemicals and Industrial Products. The individuals during the benchmarking exercise represented different parts of the business, i.e. Head Office in London, Polymers in Sunbury-on-Thames and the Research and Development Department in Hull.

Britvic Soft Drinks Ltd Part-owned by Bass, Allied and Whitbread, with links to Unilever via manufacture of Liponice. During the benchmarking visit to the regional office in Drayton House, Solihull, they were represented by the Director of Quality Management and colleagues who included the Development Director. They have been very active over the past few years introducing Quality and Process Management techniques with the aid of Xerox Quality Services, who also had a representative at the meeting.

is going; the key initiatives being undertaken; its key and core competencies etc.

9.3.6 Data collection

Each team member was equipped with a logbook to enable them to take notes during the interviews.

Once each visit was over, the members of the team involved tended to reconvene in order to complete a partner visit record using all the evidence gathered. A brief comparison between LDC's methods and practices against each of the nine partners was carried out. A sample of records filled for the items of:

- Development/innovation process structure
- Culture of innovation

is included in Appendix 9.2.

9.3.7 Closing the loop

It was extremely important for the team concerned to provide each of the nine partners involved with feedback. The team therefore undertook the task of providing a feedback report to each partner with:

- How the individual partners were perceived to be in each of the seven parameters involved
- Whether the benchmarking thought there are best practices to be highlighted and worth pursuing.

Appendix 9.3 contains an example of the standard letter sent to each of the nine partners and with the feedback report.

9.3.8 Improving the benchmarking process

Further, and in the spirit of continuous improvement, the team felt it important enough to invite views and comments from each of the partners involved on the benchmarking process itself. This proved to be a valuable exercise to have been initiated and is worth recommending to any organization contemplating benchmarking for the first time or even being experienced. The improvement in the benchmarking process means that true capability can be built-in and the teams involved can facilitate the process of developing a corporate benchmarking culture. Appendix 9.4 illustrates the questionnaire format used by the LDC team.

9.3.9 Key Findings

The scores given to each individual partner were compiled in a benchmarking study record constituted of two main parts:

1. Quantitative analysis (scores) – see Tables 9.3 and 9.4
2. Qualitative information – highlighting key areas of best practice and useful take-outs (see Appendix 9.2).

In all a matrix of benchmarks was put together (Table 9.6), highlighting the profile of each of the ten participants against the seven parameters investigated.

Table 9.7 represents the summary of classification of each partner for the seven parameters and Table 9.8 the possible follow-up that the benchmarking team have isolated as essentials or desirables of further investigation.

9.4 Action plan

This benchmarking project has demonstrated that the potential for improvement is significant and covers all the seven parameters involved, for example in the area of:

Team working: Action is:
- To provide project management and team leadership training
- Building some means of giving early warning of training needs
- Explore team-empowerment mechanisms.

Culture of innovation: Action is to have:
- A system needs to be set up to stimulate generation and allow capture, examination and selection of new ideas
- A system of individual and team awards could be used in conjunction with the above
- Idea-generation techniques such as innovation competitions, entrepreneurial play may be useful.

The outcomes of Tables 9.6 and 9.7 are, of course, useful in deciding where to further probe each item of the action plan.

In all, the LDC benchmarking project team came to the conclusion that in order to implement effectively all the areas of improvement highlighted, it will

Table 9.6 *Scores (as a percentage of maximum) given to IDC and partner companies for each of the seven parameters of the innovation process evaluated*

	Process structure	*Resourcing*	*Tools and techniques*	*Team Working*	*Documentation*	*Measures*	*Culture of innovation*	*Overall*
J	40	47	50	70	40	20	3	37
I	70	57	57	65	52	35	27	51
G	83	68	**89**	83	58	72	50	70
H	82	75	75	85	70	80	70	76
F	72	53	71	71	67	62	50	63
E	74	77	59	71	47	72	37	61
D	**89**	**83**	**94**	**88**	**90**	**88**	**87**	**88**
C	84	**83**	78	**92**	**87**	80	53	79
B	51	47	35	70	47	60	43	50
A	**91**	66	78	76	**87**	72	42	72
Average	74	66	69	77	64	64	46	65

Scores in bold print are those considered best in class.

be important to continue having a regular dialogue with each of the nine partners concerned and to use an iterative, on-going process of implementing change, recalibrating the scores and ensuring that the LDC's mission of becoming a world-class development centre in every sense can be realized.

9.5 Summary of benchmarking project management

This valuable experience of the LDC benchmarking project team was a result of thorough planning, discipline, preparedness, motivation and commitment and a broad and confident perspective of the total business. In all, it took perhaps over 15 months of hard work to get to the stage described in this chapter. What has perhaps helped facilitate the progress achieved was:

- Attention given to team dynamics
- Complementary role of team members and their valued contribution
- Good team leadership
- Good project sponsorship
- Regular communication and updates on progress
- Not being distracted by resource issue and other problems
- Willingness to succeed.

In addition to the valuable knowledge base developed as a result of this benchmarking competition, the team concerned have managed to put together

Table 9.7 *Summary of ratings given to benchmarked companies on basis of scores given after visits*

Used following 3-point scale:

- Best in class
- Better than the benchmarked average
- Average or below

	Best in class	*Better than average*	*Average or below*
1 Process structure	D A	G H C	I F E B
2 Resourcing	D C	H E	I G F B A
3 Tools and techniques	D C	H C A	I F E B
4 Team working	D C	G H	I F E B A
5 Documentation	D C A	H	I G F E B
6 Measures	D	G H E C A	I F B
7 Culture of Innovation	D	H G	I F E B A

D was considered best in class overall and for all seven parameters (jointly with other companies for five parameters).
C was equal best in class for three parameters (Resourcing, Team working and Documentation).
A was equal best in class for two parameters (Process structure and Documentation).
G was equal best in class for one parameter (Tools and techniques).
H was consistently better than average.
E performance varied between better than average to average and below.
I, F and B were consistently average or below.

Table 9.8 *Summary of need to make follow-up contact/visit to companies against the parameters benchmarked, with indication of subject of interest*

	Essential	*Desirable*
1 Process structure	B (basic design/simplified funnel)	G (pre-ideas phase) E (APP) C (cascaded missions) A (cascaded missions and project definition)
2 Resourcing	E (% innovation/time tracking) B (% innovation)	I (APP/time tracking) C (contracting out)
3 Tools and techniques	G (QFD)	E (TSP and MS project) D (MS project) A (self-assessment)
4 Team working	C (empowerment of teams and early warning of training needs)	G (Project man. and team leadership training) H (team leadership training) D (project man. training)
5 Documentation	B (auto. capture and archiving)	H (define standards) C (continuous updating) A (unify documentation)
6 Measures		E (APP) C (team feedback)
7 Culture of innovation		G (Henry Ford prize) H (supplier awards) E (Entrepreneurial Play) D (team awards, supplier awards) A (no ducks)

a total package, a tool kit for other teams to use when planning their benchmarking projects.

Appendix contains an outline of the whole process used to manage the benchmarking project in the context of the LDC. It is definitely a best practice approach that every organization wanting its benchmarking projects to result in successful outcomes should use.

Acknowledgements

The author is extremely grateful to the LDC Benchmarking Project Team (Trevor Francis, Alan Barber, Dave Clarke, Brian Humfress, Alyson King,

Denise Lofthouse, Jane Wise, Roy Palmer) for such an impressive contribution to making this project succeed. The work described in this chapter reflects the team's systematic, thorough, pioneering and effective approach to benchmarking project management.

Appendix 9.1: Questionnaire used during benchmarking of development/innovation process

1 DEVELOPMENT/INNOVATION PROCESS STRUCTURE

(a) How do you define innovation?

- Have you a mission statement which includes innovation? And CSFs etc.?
- If not, how important is innovation to your overall mission statement?
- How did you generate your mission statement and how is it communicated?
- Is innovation both consumer or technology defined?
- Have you a strategy for innovation?
- Can they tell it to us?

(b) Do you use a process management system?

- When did you introduce it?
- What have been the tangible benefits since its introduction?
- What is the main focus of the system? Time to market, quality, design for manufacturing/ assembly/in-service maintainability?
- If so, how is the process broken down?
- How do you move through the process?
- Do you have decision points?
- If so, what form do they take?
- How many stages are there in your process?
- How would you describe each stage?
- Do you have a person responsible for moving from one stage to another?
- If so, who would that be for each stage of the process?
- What are the core strengths and weaknesses of your system?

(c) What do you have to achieve to move through the process?

- Is there a checklist in use?

(d) How do you control the process?

- Is responsibility defined for each individual project, is a set protocol used?
- How do you involve your customers?

(e) Where do the ideas come from and how do they become a project?

- How do you define a project?
- How do you select and evaluate ideas?
- How do you select and monitor projects?

(f) How much time is spent on each part of a project?

- Do you use a network system when setting up your projects?
- If so, how does it work?
- Is it project size dependent?
- How much time do you spend on incremental changes?
- How much time do you spend on step changes?
- How much time do you spend on breakthrough changes?

2 RESOURCING

(a) How do you prioritize projects?

- How do you allocate resource to a project?
- Is there a limit of resource to your projects?
- Do all functions get involved in the early part of the project or only as work progresses?
- Do they still have functions? If not, how is functional expertise maintained?

(b) Describe the process for allocating resource to projects

- Is there resource from all functions within your organization?

(c) What is the proportion of funding for innovation and how is it determined?

- Are there cost/benefit techniques used to justify projects?
- Are there budget limitations allocated to projects?
- Are there time limitations allocated to projects?

(d) Is there an overall strategy for innovation?

- How is the innovation funding managed?
- State the proportion of innovation based on long-term goals
- State the proportion of innovation based on short-term goals

(e) Indicate time spent on the following activities

- Creativity
- Innovation projects
- Support projects
- Support for past projects (firefighting)

3 TOOLS AND TECHNIQUES

(a) What project management tools and techniques do you use?

Prototype facilities
Standard computer packages
Statistical analysis
SPC – Statistical Process Control
TQM – Total Quality Management
Superproject
CAD/CAM – Computer Aided Design/Computer Aided Manufacture
Spreadsheets
Virtual Reality
Stereo Lithography
QFD – Quality Function Deployment
Product/Technology mapping

What training is provided for these techniques?

(b) How do you design quality into innovation?

- How do you define quality?
- What are the key quality aspects you design in?
- Are the quality aspects built in at the beginning?
- Do you use an integrated approach?
- How do you design for manufacturability?
- How do you plan for right-first-time manufacture?
- How do you manage parallel working?

(c) How do you assess consumer/customer needs?

- How do you integrate the technical and the consumer needs into a project?
- How early?
- How frequently?
- How and when do you confirm consumer acceptability?

(d) Do you use any other quality tools?

---

4 TEAM WORKING

(a) What do you understand about team working?

(b) How do you select team members?

- Who selects the team?
- Do you use a multifunctional team?
- If so, what type of functions are involved?
- What is the structure of your teams (management: non-management)?
- Do you base teams on expertise?
- What is the average size of your teams?
- Is there a project leader?
- How is he/she chosen?
- What is the specific role of the leader?
- Does he/she remain throughout the project?
- Does the leadership/team membership change?
- Do they train project leaders and team members respectively in their roles?
- How often does the team meet?
- Are their clear roles and responsibilities (team management process)?
- How do you reconcile line/project responsibilities?
- What are the communication routes to the team (internal/external to team)?
- Have you a 'communication map' showing all the communication routes?
- How many project teams can an individual be on?
- What happens at the end of a project?

(c) What is the degree of autonomy in making decisions on a project?

- How are teams managed (internally and externally)?
- Is this consistent between teams/projects?
- To whom does the team report?

5 DOCUMENTATION

(a) What do you document during a project?

- How do you document
  Meetings?
  Results of development programme?
  Decisions?
  Changes to network?
  Results of audit?
- Do you have a standard format for a project brief?
- Who is responsible for writing the brief?
- How is it implemented/interpreted?
- Do you document fallback options within the project?
- Does the brief ever change within a project?
- If so, what changes would be made, how would they be made and who is responsible for implementing changes?
- Do you report during a project?
- Is yes, would you report and at what stages would you report?
- Who would you report to?

(b) What documentation is required to move through the project (from one stage to another)?

- Where would that documentation go?

(c) Do you have an archiving system?

- If yes what form is it in?
- Is it electronic or paper based?
- What information would be included within this system?

6 MEASURES

(a) What do you measure throughout a project?

- Time to move through the process?
- Amount of resource required for a project?
- Cost of a project?
- Quality within a project?
- Success rate of a project?

(b) Where do you measure throughout a project life?

- Do you measure continuously throughout a project?
- Do you measure at the stages/gates?
- Do you only measure at the end of a project?

(c) How do you measure?

- Does the project brief contain performance measures?
- Do you measure critical success factors?
- How is innovation performance measured against past programme, i.e. achieved against target?

(d) Why do you measure?

- What are the tangible benefits of measuring?

7 CULTURE OF INNOVATION

(a) Do you have incentive schemes (innovation prizes)?

(b) What do you understand by a culture of innovation?

- In your own view, have they achieved it?
- Do you allocate time for creativity?
- Do you have creativity sessions?
- Do you allocate time for creativity?
- How do you produce/capture ideas?
- Have you got suggestion schemes for innovation (ideas-generation process)?
- Do you encourage employees in external innovation activities?
- Do you have a forum for discussion of ideas? (Ours is singular championing)
- Can anyone contribute input/ideas?
- What relationship is fostered with suppliers/partners?
- Is it a key part of their innovation programme?

(c) Do you have an innovation team?

- Is there a senior innovation champion?
- What % of employees are involved in external innovation activities?
- Do you audit innovation culture?
- How is this accomplished?
- Do you manage innovation reactively or proactively?
- Are you committed to partnerships for supplying innovation activity?
- What methodologies do the teams use to generate new product ideas?

(d) What is use/attitude to patents?

Appendix 9.2: Benchmarking study record: summary of conclusions and actions after visits to nine partners

1 Development/innovation process structure

LDC method or practice	*The benchmarking method or practice*
In principle the Innovation Funnel is used. In practice, the comprehensive documentation and clear decision points required to move through the process are in place. However, it is expected that in the near future the Integral APP methodology will be implemented, thus formalizing the practice.	**A** possess a mission statement which is cascaded and made relevant for each function. They have had a NPD process in use for nearly 3 years, similar to a funnel with a gate management process, and with documented deliverables required at each phase. Have a project definition ('Non-routine action to meet specific needs'). A checklist for managing projects (major and minor) is common throughout the company. NPD process starts with a strategy involving market research department and a NPD team (marketing, technical, operations, etc.) right from start. Enables innovation to be both consumer and technology driven. **D** have had innovation funnel/gate system in place for 5 years, software used to lead team through system (Commercialization Process). All teams empowered. Mission reinforced through display on PCs when switched on. Ideas collection is a full-time job. Portfolio of ideas v. used. Ideas are archived and can be easily recalled. All documentation for a project can be traced back electronically. Checklists used. Any opportunity must pass basic criteria to move through system. If project is in a new area then a special team is set up. Team decides which are the best opportunities. Milestones set by team members. Regular reviews at milestones to audit progress.

Potentially useful take-outs

ex A. Should consider the principle of defining innovation strategy through involvement of a NPD team and market research. The need to present an annual NPD strategy to obtain budget approval encourages proactiveness.

ex D. Mission statement on PCs when switched on. Ideas collection as a full-time job. Ideas archiving system (1200 in system at present).

Potentially useful take-outs from other partners

ex G. The principle of feedback from a post-launch monitoring phase into the ideas phase needs to be given more emphasis by LDC (and Unilever).

ex H. Reliance on plan/do/check/act regime provides a good discipline. Putting effort into continuously improving how things are done as an integral part of the project (rather than just improving the product) has long-term advantages.

ex I. and E. Implementation of the integral methodology should enhance awareness and practice of project management tools and techniques.

ex F. Have a system very similar to integral.

ex C. Range of mission statements to suit different levels. Don't forget to keep updating process. Top-down demand to use/improve process.

7 Culture of innovation

LDC method or practice	*The benchmarking method or practice*
At present little official development and support of a culture of innovation is evident. Incentive schemes have been fragmented and often only lasted one year.	**D** used to give people a card and candle (for a Bright Idea). Idea generation is automatically in the project team. There is a commercialization team award, i.e. team is rewarded. Occasionally have brainstorming. 80% of ideas from customers, but person who brings them into business is acknowledged as the generator. Trying to provide people with 'free' time to think and invent. No one person can kill an idea, only the team. Ideas collection and archiving is a full-time job. Patents encouraged but now more selectively.
Potentially useful take-outs Team awards have benefits in encouraging teamwork.	

Potentially useful take-outs from other partners

ex H. It will clearly be beneficial to become more involved in encouraging all to put forward potentially useful ideas, and in providing rewards for those whose ideas prove useful. The possibilities of providing awards for good performance in a variety of other activities need to be explored.

ex G. Motivational benefits of a continuous and consistently applied recognition scheme are claimed.

ex F. Get an idea approved? Go and do it!

ex E. Having and implementing a strategy for proportion of effort on ideas generation would be useful.
Formalization of the ideas generation process should have benefits.
Entrepreneurial play seems useful.

ex C. Interteam exchange of information/progress/developments.

ex B. Innovation competition?

ex A. Key learning is the ability to recognize/reward good ideas and work at any level.

Companies have been disguised to protect sensitivities.

Appendix 9.3

Response form for visit to I

Parameter	*Comments*	*Lever follow-up*
Process structure	Average or below	None
Resourcing	Average or below	Desirable
Tools and techniques	Average or below	None
Team working	Average or below	None
Documentation	Average or below	None
Measures	Average or below	None
Culture of innovation	Average or below	None
Overall	Average or below	-

Comments: Best in class
Better than the benchmarked average
Average or below

Follow-ups: Essential
Desirable
None

List of benchmarked topics: definition of aspects of parameters being evaluated

Benchmarking is the process of finding and implementing best practices. (Requires being humble enough to admit that own practices need to be improved and we can learn from others.)

The LDC benchmarking team is benchmarking the development/innovation process using the following parameters:

1. **Development/innovation process structure**
   How well defined and easy to communicate/understand/utilize is the structure being used? How rigorously is it applied and controlled?
2. **Resourcing**
   How effectively are projects prioritized and the required cross-functional resources calculated, deployed and monitored?
3. **Tools and techniques**
   How comprehensive is the range of techniques available for all aspects of the innovation process and how effectively are they used?
4. **Team working**
   What criteria are used to select team leaders and members and how is effective empowerment and cross-functional cooperation achieved? What training is given in project leadership and team working?
5. **Documentation**
   How well documented and archived are the critical stages of the innovation process, so that the control stages (e.g. briefs, charters, CSFs at gates) are rigorously applied, checked and recorded? What specific systems are used?
6. **Measures**
   What measures are made before, during and after the innovation process and how do they assist in improving the effectiveness of the process? What review mechanisms exist for the measures?
7. **Culture of innovation**
   How and how well does the organization promote a culture of innovation and encourage all to contribute towards it by appropriate organizational devices and rewards?

Standard letter sent to each of the nine partners

Dear

BENCHMARKING THE INNOVATION PROCESS – FEEDBACK

We have now reached the end of our initial series of visits for our project and would like to thank you once again for your help.

Now comes the difficult task of building our findings into implementable actions. However, at this stage, we felt the need to provide a little feedback to you on what we have found and solicit your comments on how you have seen our project.

Enclosed with this letter you will find:

(a) **List of benchmarked topics**
This shows the seven areas we have assessed with a brief description of the aspects of interest to us.

(b) **Our rating of your company**
Please bear in mind that the assessments we have made are necessarily highly subjective and based on our own particular interests and requirements.
We have used a very simple three-point scale for comparative assessment and have often accepted more than one 'Best in Class'.
This summary also includes the team's internal recommendations on the value of follow-up contacts (should this be mutually acceptable) when we begin to deal with the practicalities of implementation.

(c) **Process questionnaire**
We would very much appreciate your responses to this simple questionnaire, especially as this is the first exercise of this type that we have undertaken. Your responses can be made anonymously if preferred.

As you will see, Unilever faces some pretty tough competition in the areas we have been benchmarking. However, we are sure we will benefit from further contacts as we begin to implement APP (e.g. your experiences with time tracking) and as we begin to explore the potential of QFD.

In the meantime, if you have any queries or would like further background on the ratings we have generated, please feel free to contact any of the team.

Thanks again for your assistance.

Yours sincerely,

Appendix 9.4: LDC Benchmarking team Process questionnaire

Name (optional): ....

Company (optional): ....

1 Spontaneous comments

Was there anything you particularly *liked* about the way we carried out this exercise?

Was there anything you particularly *disliked* about the way we carried out this exercise?

2 Specific ratings

Please answer the following questions by ticking the appropriate columns.

	Excellent	*Very good*	*Good*	*Mediocre*	*Poor*
Pre-meeting					
Clarity of initial contact and background given	( )	( )	( )	( )	( )
Follow-up paperwork and expansion of topics to be covered	( )	( )	( )	( )	( )
Handling of meeting arrangements and confirmations at Lever end	( )	( )	( )	( )	( )
Meeting					
Clarity and relevance of introductory material	( )	( )	( )	( )	( )
Clarity and professionalism of discussions	( )	( )	( )	( )	( )
Matching of time allocated with agenda points to be covered	( )	( )	( )	( )	( )
Coverage of all relevant points	( )	( )	( )	( )	( )
Size and composition of team	( )	( )	( )	( )	( )
How appreciative were the team of your time and efforts	( )	( )	( )	( )	( )
Post-meeting					
Response to any of your queries or visit requests	( )	( )	( )	( )	( )
Speed of post-meeting feedback	( )	( )	( )	( )	( )
Quality and depth of post-meeting feedback	( )	( )	( )	( )	( )

3 General improvements

Are there any specific improvements you would suggest?

Any other comments?

Index